Data Analysis

An Introduction

Data Analysis

An Introduction

Bryan Nolan

Polity Press

First Published in 1994 by Polity Press
in association with Blackwell Publishers

Editorial office:
Polity Press
65 Bridge Street
Cambridge CB2 1UR, UK

Marketing and production:
Blackwell Publishers
108 Cowley Road
Oxford OX4 1JF, UK

238 Main Street
Cambridge, MA 02142, USA

ISBN 0 7456 1145 1
ISBN 0 7456 1146 X (pbk)

A CIP catalogue record for this book is available from the British Library and the Library of Congress

Typeset in 11 on 13 pt Times
by TecSet Ltd
Printed in Great Britain by T.J. Press, Padstow, Cornwall

This book is printed on acid-free paper.

Contents

Author's Note

1 Introduction and Overview

This book is an introductory text in data analysis designed for students on the first year of a social science degree. The examples I have chosen have been drawn from various areas of social science. However, I have tried to avoid the specialized language or jargon of these disciplines as much as possible and to describe concepts in simple terms.

The main focus is, of course, on the area or discipline of data analysis and as such the text should also be a useful introduction to data analysis for students on business, health care, social policy and other courses.

We might ask ourselves at this point: What is data analysis? The whole of this short text is an attempt to give you an understanding of the techniques and concepts of data analysis and this is a question you should be able to answer more adequately when we arrive at the end of the book. However, as an initial or working definition we could say that data analysis is a way of making sense of the patterns that are in or can be imposed on sets of figures.

The techniques are generally drawn from statistics and exploratory data analysis – a form of elementary but effective analysis developed by John Tukey and his associates. As 'tools' we shall use a pocket calculator and a computer package called Minitab.

I have tried, where possible, to use small data sets to introduce various concepts. I am firmly of the opinion that you cannot understand the various principles and techniques of data analysis unless you start with relatively small data sets and initially analyse them 'by hand'. That is, in our first approach we shall use pencil and paper and then a pocket calculator. Sometimes this can be tedious, but it is necessary tedium. Only when we have gained a reasonable understanding of the technique and its limitations should we use the computer to help us. Computers will only do what we tell them to do - and sometimes it is hard to get them to do this! They have no true discrimination of their own, so we need to use them as tools to *help* us in *our* analysis. That is, they cannot do the analysis except under our direction, and to direct them effectively we need to have a reasonable idea of what is going on in the first place. Hence our initial approaches to analysis will be modest and by hand. Later we shall use the computer to remove the tedium from the act of analysis.

As this book is a simple introduction to data analysis I have adopted the 'bottom-up' approach described above. In my experience this is an effective way for students to gain a foothold in the field of data analysis.

The initial use of small data sets, which are often fragments of real data sets, seems to be the best way to approach the subject. For instance the fundamental concepts of average and spread, which are introduced in chapter 4, are most easily grasped using small and manageable data sets and a pen–paper–calculator approach. Again the rather more difficult concepts that lie behind the fitting of a straight line to plotted data points (chapter 7) are also usefully approached using small data sets. As I hope you will find, such an approach using a small set of numbers gives a 'feel' and understanding of the processes of line fitting. Such an introduction to line fitting should also prevent you being tempted into the 'push button' approach to the calculation of regression and other line-fits which is so easy with modern computers.

Small fragments of data sets are also used initially in chapter 2 to introduce you to the Minitab package. You will find that, in this and other chapters, concepts are generally introduced using small data sets; however, some larger data sets are also used such as the ones on pulse rates, suicide, and the UNICEF economic and health data.

This is an appropriate place to mention some of the dangers of computer packages.

We are, these days, told by the government that cigarettes can damage our health in various ways. In a similar fashion we should also be aware that computer packages can seriously damage our understanding. It does not take long to learn to use a computer package. Minitab, like some others, is very user friendly, having good error messages, a variety of prompts and comprehensive help programs, and consequently we can soon learn to feed the data into the computer and 'crunch' the data in a variety of complex ways.

Here lie both the power and the danger of using computers for analysis. Of the vast array of techniques available in the package, have we chosen the ones that are appropriate for this particular data set? If the type of analysis is appropriate and we have not made any mistakes, how can we understand and interpret the findings? Are such findings significant or even meaningful?

To begin to answer these difficult questions we must have a reasonable knowledge of both data analysis as a subject and the area of study concerned, say psychology, demography, medicine, nursing etc., from which the original data were drawn.

In data analysis today it is fatally easy to 'crunch' vast amounts of data in complicated ways and end up with lots of figures and 'findings' which can be beautifully presented in graphical and other forms which, despite their appearance, are meaningless. You should stick with simple forms of analysis – 'by hand' initially – until you have developed a reasonable understanding of the techniques outlined in this introductory text.

Note that I say a 'reasonable' understanding. A complete understanding of the concepts and techniques of data analysis or statistics can, in my view, only be achieved by studying them thoroughly but in a preliminary way and then using them. Full understanding rarely, if ever, comes from a study of concepts and first principles, but from a sustained attempt to understand the principles followed by a tentative or exploratory *use* of the techniques.

Understanding in this sense is a type of dialectic or twofold process; it is analogous to developing hand and eye co-ordination in a sport. The principles and concepts need to be appreciated but understanding is only achieved by attempting to use the techniques while continually referring to the underlying principles and concepts. We need both theory and practice and the working understanding that comes from the dialogue between the two.

In the context of this necessary dialogue between theory and practice it is important to remember to think about and comment on the analysis as it proceeds.

It is worth developing the habit of initially analysing the data in a small-scale and tentative fashion and writing a short critical commentary at each stage of the analysis. Analysis, commentary and the written word have a place not only at the end of numerical analysis but also throughout it. Building up a critical commentary as we work through the data makes the task of writing the final summary and conclusion much easier. I am pointing to the need for each analysis to be concluded by words. As my teacher in primary school used to say, in a different context, 'After SUMS, we will do WRITING'.

What I am bringing to your attention here is that serious and sustained thinking about data in terms of the concepts of data analysis does not end with the production of numerical summaries and significance levels, but with words.

It is worth noting that well-organized researchers keep detailed journals or notebooks of successive analyses done by hand and by computer and always comment on computer print-outs as they proceed through an analysis or research project.

Outline of the Chapters

Chapter 2 The Minitab Package

This is a warming-up, hands-on chapter and in it I introduce you to the Minitab package. It is laid out in a series of graduated exercises or tutorials. The aim here is not to introduce any of the concepts of data analysis but merely to give you some familiarity with the computer package which will be used in subsequent chapters.

Chapter 3 Looking at Data

Using two simple data sets of the heights of 100 males and 100 females I look at various ways of organizing and presenting the data so that we can best detect any pattern in it. Measures of average are introduced along with classes, histograms, frequency polygons and stem-and-leaf diagrams.

Chapter 4 Level and Spread

Here I look at two types of measure or numerical summary which can help us describe and understand a set of data:

1 measures of level or average;
2 measures of spread or dispersion.

The sigma notation is briefly outlined.

Chapter 5 Exploratory Data Analysis of Two Data Sets

In this chapter I illustrate some of the techniques outlined in the previous chapter by looking at two data sets, namely the heights and weights of 45 men and 39 women and then the age and sex distribution of suicide in 16 countries (abstracted from *World Health Statistics Annual 1988*).

Chapter 6 Two Variables, Correlation, Cause and Effect

Here I look at some ideas of the relationship between two variables, using initially the related variables of height and weight of 60 seamen. I examine the notion of causality before outlining a simple but useful notion of correlation.

Chapter 7 Fitting a Straight Line to Data

The least-squares or regression method of line fitting is introduced. We first use a small and manageable subset of the heights and weights looked at in the previous chapter and then we utilize Minitab to help us with larger data sets.

Chapter 8 Linear Transformations

In this chapter I outline some simple ways to 'unbend' data by transformation so that we can extend the use of the powerful technique of line fitting outlined in the last chapter.

Chapter 9 The Gaussian or Normal Distribution

After looking briefly at the pure or mathematical form of the normal distribution and saying a few words about its discovery or invention I turn to three types of distribution which can approximate to the normal:

1 heights and weights in the 'real' world.
2 binomial probabilities.
3 sampling distributions.

These distributions are illustrated by computer simulations before we finally look at confidence intervals for population means.

Chapter 10 Using Standard Normal Tables

Here I pick up the topic of the standardization of data first outlined in chapter 5. Normal and near-normal distributions are transformed to the standard normal form and probabilities are related to the areas under the curve. Tests of hypotheses about the population mean are outlined.

Chapter 11 The Chi-Square Test

In this final chapter I look at the chi-square distribution and say a bit about its relation to the normal distribution and how we can simulate a sampling distribution of chi-square. I then look at hypothesis testing using this distribution and end by looking at contingency tables.

Each of the chapters above has exercises and answers for you to work through.

How to Read and Use the Book

The first point to make is the obvious one that it is not possible, with the exception of this present chapter, just to read through the book. It must of course be *worked* through. That is, you should do for yourself both the worked examples and the exercises at the end of each chapter. This idea of 'working through' examples – even if they appear to be straightforward and obvious – is a very important one in data analysis, and indeed in all mathematics.

It is generally only when we work an example through for ourselves that we really make the knowledge our own. It is a common experience amongst students of mathematics that they sit through a lecture which seems to them to be crystal clear and indeed sometimes boringly obvious and yet are unable to understand the work later until they have worked through the material.

In this context it is also worth mentioning that helping or teaching other students is beneficial. Working on data analysis in small groups, helping each other and discussing the work is an excellent way to learn. It helps both the weak and the strong student. Weaker students can benefit from the support of the group and stronger students can learn and consolidate knowledge by helping others. This is not a matter of altruism, most people find that articulating and explaining a problem for someone else is an excellent way to understanding.

I have aimed to write a short coherent introduction to data analysis techniques and unless you have some prior knowledge of the area the place to start the text is at the beginning, i.e. with chapter 1. You should then work through each chapter in turn being careful to complete and keep notes on the exercises.

However, if you want an even shorter introduction you could, if you wish, miss out the linked chapters 7 and 8 at a first reading. These two chapters deal with fitting straight lines to data and transforming data and subsequent chapters do not depend on them. You can always come back to these two chapters later. They do, however, contain important concepts and techniques which are extremely powerful and useful and which lend themselves admirably to computer analysis.

In order to spare you the tedium of making up data sets I have gathered together on disk a number of small data files. These files which are referred to in the text are labelled DATA01.DAT, DATA02.DAT etc. Should you wish, the disk can be purchased from the ESRC Data Archive at the University of Essex for £12.50 ($19). Cheques should be made payable to the University of Essex and you should indicate whether diskettes (3.5" only) are to be used in an Apple MacIntosh or MS-DOS environment. The address and telephone number is as follows:

ESRC Data Archive,
The University of Essex,
Colchester,
Essex CO4 3SQ
England
Telephone: 0206 872001 (UK)
011-44-206-872001 (from USA)

So that this introductory chapter is not totally given over to chat about data analysis let us end by looking at three substantive and important concepts in statistics and data analysis: population, sample; and variable.

Some Working Definitions

Let us look at some simple initial definitions of the important concepts of population, sample and variable that we shall encounter later. What are given here are to be understood as working definitions, definitions which will enable us to begin to understand the concepts and so follow their development through the book. By the end of the text we will have developed a much greater understanding of the basic concepts described by these terms.

Population

The population, or universe as it is sometimes called, is the larger set or group of things or events being studied. The term reminds us that organized data gathering was originally about human head-counting, a state arithmetic to determine manpower and other resources for warfare or economic reasons. However, in data analysis today the term is used to indicate all kinds of groups or sets of things as well as people.

It is important to note that population is a defined thing, i.e. it is defined by the person collecting the data. This type of population, unlike more naturally occurring populations such as a population of people, is a human artifact, and is best thought of as the whole of the set of things that we wish to study. As such we must define what will be included in the population and what will not. A population has stated boundaries and is an aggregate or set of individual units or elements of a defined type. We can for example have a population of the heights of all children in a given school, or a population of the weights of all schoolgirls of age 6 years in the Borough of Greenwich in 1945, or a population of the circumference of the heads of day-old babies born in a given London hospital during August last year, or a population of the breaking strains of 100 steel cables, or a population of all babies under 2500 g weight born in England and Wales today.

These are finite empirical populations. Finite populations are those populations which contain a precise number of elements which, at least in theory, can be counted. Other types of population can be generated or derived mathematically, as we shall see in later chapters, but we will start by looking at finite empirical populations such as the ones above.

Each population will have certain features or characteristics which may be found or estimated. We may, for instance, be able to calculate some kind of middle or average value for the data in the population, or some measure of spread or dispersion of the data in the population about this central point. We would then have two measures that relate to the population, a measure of average and a measure of spread. These, and many other useful measures which describe a population, are called population parameters.

Sample

A sample is a subset of the data in the population. It is a part of the population that has been chosen from it. There are of course many ways of selecting subsets from a population, but we will only interest ourselves in those types of samples which can represent the whole. Only certain methods of selection ensure that the sample is representative. Taking a sample is only worthwhile if we can use it to say valid things about the parent population from which it was drawn. That is, we are only concerned with representative samples.

We are extremely fortunate that such representative samples can be formed in a direct and easy way. This involves choosing elements for the sample from the population in a certain very precise way. This is called random sampling, and we will of course examine it in more detail later, but it needs to be stressed here that, contrary to the way the word random is sometimes used in everyday speech, there is nothing haphazard, casual or slap-dash about the choice of elements in random sampling. Here, random sampling has an exact meaning, and in the simplest case it means that each element in the population has an equal chance of being chosen for the sample. Substituting the word probability for chance we can see that it is a type of probability sampling. With repetition such individual choices or probabilities can generate stable patterns or probability distributions.

Having drawn a sample of data from a population we may then analyse it and calculate, say, measures of average and spread. These and other measures which describe the sample are called sample statistics. However, our interest does not stop with such sample statistics, or even with the sample itself; what we are really interested in is how such characteristics of the sample relate to the corresponding characteristics of the population from which the sample was drawn. What, for instance, is the connection between the sample average and the population average? How is the spread found in the sample related to the spread of the population data? That is, we wish to infer things about the population as a whole from the knowledge we have gained from the sample. We wish to convert our sample statistics in some way into reliable population parameters and so gain some knowledge of the parent population.

This process is called statistical inference. It is to do with drawing valid conclusions about a whole from a study of its parts. The validity and accuracy of such inference depends on many factors, some of which are obvious, such as the size of the sample. As one would expect, the larger the sample, the better the inference concerning the population is likely to be. However, it is worth stressing again that the crucial factor underpinning such inference is whether the sample was a random one. The validity of the inference depends on the method by which the sample was chosen.

Variable

The variable is the thing that varies. It is the individual element, characteristic or unit that makes up the data in the population, and of course the sample. It is the thing we set out to collect and measure or count and we usually give it the symbol X. The variable will have a dimension attached to it, such as the dimensions of weight, time, length, dollars etc.

Variables are of two general types, discrete and continuous. Discrete or counting variables are those variables that vary in discrete or unit jumps, such as, for instance, the number of people in a family, or the number of planes landing at Gatwick on a given day. A continuous or measurement variable is one where increments of less than a unit are possible, and, given the need and the instruments, we can measure each element to any degree of accuracy we wish. Height, age and weight are examples of continuous variables as, theoretically at least, we can obtain a third measurement between any two adjacent measurements. The appropriate method of collection for discrete variables is counting, and for continuous variables is measurement. In the case of measurement the degree of accuracy needs to be stated. For example, weights should be stated to be measured to the nearest kilogram, salary measured to the nearest dollar, cent etc.

Short Bibliography

The following books are of two kinds; those that I refer to in the text, and books on elementary data analysis and statistics that you could consult if you wish to extend your study of the subject matter introduced in the following chapters.

The first three texts relating exclusively to Minitab can be purchased from Clecom, The Research Park, Vincent Drive, Edgbaston, Birmingham B15 2SQ, England or Minitab Inc, 3081 Enterprise Drive, State College, PA 16801–3008, USA.

Minitab Reference Manual, Release 8, PC Version. Minitab Inc., Pennsylvania State College, PA, 1991

Minitab Quick Reference, Release 8, PC Version. Minitab Inc., Pennsylvania State College, PA, 1991.

Minitab Handbook, Ryan, B.F. Joiner, B.L. and Ryan, T.A. PWS-Kent, Boston, MA, 1985.

Understanding Data, Erikson, B. and Nosanchuk, T.A. Open University Press, Milton Keynes, 1992.

Elementary Statistics Tables, Neave, Henry. Routledge, London, 1992.

Fundamental Statistics for the Behavioral Sciences, Howell, David. Duxbury, Boston, MA, 1985.

Statistics without Tears, Rowntree, Derek. Penguin, Harmondsworth, 1991.

Exploring Data, Marsh, Catherine. Polity, Cambridge, 1988.

2 The Minitab Package

This is a warming-up or hands-on chapter and it is devoted to helping you understand the Minitab package. In it I give a brief overview of Minitab and then guide you step by step through various exercises using two command systems to operate the package, Menu Commands and Direct Commands.

My intention in this chapter is *not* to teach you data analysis – that is reserved for subsequent chapters – but to guide you through various parts of the Minitab package so that you can gain confidence with it and establish a sufficient foothold in it to appreciate the text of the following chapters, where the focus will indeed be on data analysis rather than on the computing element.

After having read the introduction to the parts of Minitab you should start on the exercises. The best way of tackling them is, first to read the exercise through to get a general idea of its aim, and then to start working it through on the computer. It is only by doing the exercises yourself that you can learn the rudiments of the package. Reading alone is quite insufficient.

Probably, by the time you get through the first three or four of these eight exercises you will be at the threshold of an appreciation of the main features of Minitab and could start working through other chapters. However, you may wish to consolidate your knowledge of Minitab with the remaining exercises in this chapter before going on.

In working through these exercises, despite care and attention, it is probable that you will get stuck in places. If this does happen, you should note that I have given the Menu Commands in sections so that you can use the Escape key to get out of the menus and start the section again. If you still find yourself sticking it is probably worth going on to the next section of the exercise if no help is to hand. You can always return to a sticking point at a later date.

In this context it is worth noting that even experts in computing get stuck sometimes, especially when they are tired or having a 'bad day'. There are at least three things that you can do when stuck. First, check the precise wording, form, punctuation, structure and spacing etc. of the command that seems to produce the wrong response. Second, carefully read any error messages that are displayed, and third, use the Help program. If none of these produces results then it is probably time to go for a cup of coffee and a break!

When you are learning to use a new computer package, a Help program can be invaluable in sorting out errors and giving guidance. Minitab has a good comprehensive Help programme that not only will aid you initially but will be invaluable as you explore more advanced areas of the package.

There are many ways to call up this Help programme, the easiest being to press the F1 key or type HELP after the Minitab prompt, i.e. MTB > HELP. Either of these commands brings up the Help menu from which you can choose the kind of help that you think you require. However, if you know what type of help you need it is easier to call for it directly. If, for instance, you want help regarding the form of a command or regarding the drawing of histograms, you should type as follows after the Minitab prompt:

```
MTB > HELP COMMANDS   or   MTB > HELP HISTOGRAMS
```

As you sit down to work through the exercises in this chapter, I am assuming that you have access to a version of Minitab and also have the use of a printer. In addition I am assuming that you have enough basic knowledge to switch the system on and off, format and handle disks, and use the keyboard and printer etc. If you have a recent version of Minitab you will be able to work through both sets of exercises that I give in this chapter, those using the Menu Commands and those using the Direct Commands. If you have a version of Minitab older than Release 8 you will need to use the Direct Commands exclusively. Although this book is written primarily for PC versions of the package it can still be used with mainframe versions; in this case, however, you will probably need to make minor adjustments to what I say regarding entering and leaving the package, saving and editing some data and utilizing printers and other peripherals. Despite small differences in PC and mainframe versions the main body of the Minitab package is substantially the same in each case. Mainframe users should have no difficulty in following the examples and exercises in this book.

Before starting on the exercises and the overview of the parts of Minitab that precede it, I will say a few words about the notation used in this text.

Alt+D, Alt+M, Alt+H etc. are two-finger commands and mean that you should hold down the Alt key whilst pressing the key indicated after the plus sign. Menu commands are given in the following form:

```
Choose Stat▸ Basic Statistics▸ Correlation▸
Text box,type Weight Age▸ OK
```

The sign ▶ indicates that you should move to the next part of the command using one of the navigation keys. These are Tab, Enter and Arrow keys and the Space Bar. The above is a single sequence of Menu Commands and it indicates that you should do the following: first go to the Menu Strip and choose the Stat option, then go to Basic Statistics, followed by Correlation. This is achieved with the Alt+S command and then navigating up and down the drop-down menus with the Arrow keys. Choices are registered or entered with the Enter key. Once we arrive in the Correlation Dialogue Box (these terms are explained later), we navigate or move around it using the Tab key. Repeated pressing of this key in Dialogue Boxes sends the cursor on a closed circuit of various elements in the box. In the command above, we move around the box with the Tab key until we come to the Text Box (explained later) where we type in the variable names, Weight and Age. The command then tells us to move to the OK, which we do with the Tab key, and when we get there we enter or register our choice, not with the Enter key this time but with the Space Bar.

Parts of the Minitab Package

Initially we can think of the Minitab package as comprising two parts: the Basic or Main Structure and alongside it the Data Editor. I will give a brief overview of these two parts, looking first at the simpler structure of the Data Editor.

Data Editor

This is the part of the package in which we enter and edit the data to be analysed. We can think of it, at this stage, as a rectangular network or matrix of rows (R1, R2, ...) and columns (C1, C2, ...) in which we can store data. Generally speaking each column will contain the data for one variable, such as weight or age, and every row in this column will contain just one observation of that variable. For instance if we have the birth-weights of 100 babies we can enter them in, say, the first column, C1, and label it Weight. Each row, R1, R2, ..., R100, of this first column, C1, will contain the weight of just one baby.

We can switch between this Data Editor and the Basic Structure of Minitab using the following commands:

Alt + D to go from the Basic Structure to the Data Editor
Alt + M to go from the Editor to the Basic Structure

Basic Structure

This is the main or central part of Minitab and consists of about 200 or so operations or groups of commands with which we can analyse, manipulate or transform the data that we have in the Data

Editor and produce various types of statistical summaries and graphs etc.

The 'surface' of this Basic Structure is viewed through what I shall call the Primary Screen (also called the Session Window in some Minitab literature). This is the first screen that you will see when you enter the Minitab package.

In its initial stage the Primary Screen has a Menu Strip across the top, instructions on how to open a menu and the Minitab prompt MTB >, and a flashing cursor. When we have this Primary Screen in front of us we can activate the 200 or so Minitab operations in one of two ways:

(a) by Menu Commands, using the Menu and Dialogue Boxes, or
(b) by Direct Commands, which are typed in the Primary Screen after the Minitab prompt. In some Minitab literature these are called Interactive or Session Commands.

As an example of a simple Direct Command we could have, say,

```
MTB > PLOT C1 against C2
```

This command would give us a graph of the data in C1 and C2 plotted against each other on simple orthogonal (right-angle) axes.

Which of these two means of communication, Menu Commands or Direct Commands, we use is a matter of individual choice (assuming that you have Minitab Version 8 or later) and depends very much on the task in hand. Users of Minitab often have a preference for one or the other type of command but it is probably a good strategy to be familiar with both. Most users end up using a mixture of both types of command in their work.

When we know exactly what type of analyses we wish to perform on the data it is usually quicker to type the Direct Commands onto the Primary Screen after the Minitab prompt. However, it is also advisable as a newcomer to the package to spend time becoming thoroughly familiar with the Menu and Dialogue Box system as it can give you a good overview of the wide range of options that the package has.

Additionally it is worth making the initial effort to learn to use the Menu Commands efficiently as a great many computer packages of all kinds are now of this menu-driven type. Once one has developed some facility with one menu set-up it can give the experience and confidence to transfer to other computer packages more easily. There are several fairly standard procedures and keys that many menu packages utilize to navigate through menu and other boxes. Amongst these keys are the Arrow, Tab, Space Bar and Enter keys that we use in Minitab.

In relation to these two modes of command, Direct and Menu, you should note that when the menu system is used Minitab 'follows' your Menu Commands by entering the Direct Commands onto the Primary Screen at the completion of the sequence. These, with the ensuing computer calculations, constitute a record of the analysis which can be filed and printed out. That is, the Menu and Direct Commands give precisely the same results in the end and the Primary Screen record is always kept in terms of the more fundamental Direct Commands.

As will be seen, it is relatively easy to guide your analysis using Direct Commands, which are relatively clear and unambiguous. It is a longer and more cumbersome task, however to talk you through comparable Menu Commands. I have attempted in this chapter to give you some instruction in the use of Menu Commands by working all the examples in two ways, first by Menu and then by Direct Commands. As mentioned earlier, I think it is desirable that you learn to use both command systems. As well as the obvious benefits that I mentioned, this will enable you to switch easily from one to the other and, for any given analysis, to make an informed choice of which set of commands and options to use.

It should be noted, however, that apart from the exercises in this present chapter, which duplicate the commands, I have focused explanations, worked examples and answers to exercises on the Direct Commands and the record of commands and analysis that we get from the Primary Screen. I have done this for brevity and in order to get concise and unambiguous results files. However, when you work through subsequent chapters of this book you may, of course, either choose the Direct Command route or use the Menu Commands. Errors aside, these will both give the same results.

Before looking at the hands-on exercises I wish to say a few words about the Menu and Dialogue Boxes. In addition to the points made below you should refer to the extensive Help Menu if you feel the need. You may also have access to a copy of the

Minitab Reference Manual, PC Version, Release 8 (November 1991), which is the definitive reference for this package. It is comprehensive but correspondingly expensive.

As a handy, but cheap, reference you may also find it useful to buy a copy of the Minitab booklet of 8 pages called *Menu Quick Reference*, Release 8, PC Version. This can be obtained from Clecom, The Research Park, Edgbaston, Birmingham B15 2SQ, England, at a cost of about £2 (in 1993), or Minitab Inc 3081 Enterprise Drive, State College, PA 16801–3008, USA, at a cost of about $3.50 (in 1993).

However, the hands-on examples in this chapter and the use of the F1 Help programme should give you an adequate and reasonable foothold in the Minitab package and provide a good base from which to follow the various analyses in subsequent chapters.

Before looking at the Menu and Dialogue Boxes it might be well to note how to start and stop Minitab.

To start Minitab Computers in many universitites and college laboratories are set up with an outer 'shell' which lists all the packages contained; if this is the case you should start Minitab as directed in the 'shell'. Otherwise go into the operating system and at the DOS prompt simply type in the word Minitab, i.e. C:\>MINITAB. This will give the title screen followed by the Primary Screen with the flashing cursor and the Minitab prompt, MTB >, etc. You can now use either Menu or Direct Commands.

To stop Minitab At the Minitab prompt, type in Stop, i.e. MTB >STOP. There are also Menu Command ways of stopping or we can use the two-finger command Alt+X.

Let us now look in greater detail at aspects of the menu set-up.

Menu Boxes

Menu Boxes are relatively straightforward to use. When you start Minitab and look at the Primary Screen you will see a Menu Strip of six elements across the top of the screen; the File, Edit, Calculate, Statistics, Graph and Help menus. We drop down the desired menu using the Alt key followed by one of the letters F, E,

C, S, G or H. For example, Alt + E (press the E key whilst holding down the Alt key) will drop down or open the Edit menu. The Help menu can be opened using either Alt + H or F1. However, you should note that once the cursor is in the Menu Strip, after using any of the two-finger commands above, it can be moved with the Arrow keys to any of the menus.

Once the desired menu is chosen and opened the Arrow keys can be used to position the cursor or highlighted bar on the required option, which can be selected by pressing the Enter key.

Menus can be accessed in this way from most, but not all, positions in an analysis session. To close a menu and return to a menu-free Primary Screen you should press the Escape key. Parts of the menu system can also be accessed using highlighted letters that appear on various lines of the menu; however, the more general method given above using the Arrow keys is probably easier for a newcomer to Minitab. You should also note that in this context all the above can be done with a mouse. Release 8 of Minitab supports the use of a mouse but does not require it. For brevity and in order not to confuse the newcomer to computing this text will stick to keyboard commands but the experienced Mouser should have no difficulty in using the corresponding Mouse Commands. Mouse Commands in Minitab follow the usual pattern of choosing various elements in Menu (and Dialogue) Boxes and clicking them. The Help programme does have a little mouse instruction and a fuller range of mouse instructions can be found in the *Minitab Reference Manual.*

When you look at the menu options you will see that some are followed by dots and others by chevrons. For instance in the File menu we find:

```
Save Worksheet as ...
Other Files         >>
```

The chevrons indicate that this selection opens a submenu and the dots indicate that choosing this menu option leads to one or more Dialogue Boxes. We shall now focus our attention on the use of these Dialogue Boxes which you will find requires more practice than simple menus.

Dialogue Boxes

Here, rather than dealing with a range of menu options, we have a variety of elements in different parts of the screen that we have to navigate amongst and register our choices. We also have boxes within boxes. There are elements of four main types within Dialogue Boxes: List Boxes, Text Boxes, Check Boxes and Buttons. There is unfortunately no such thing as a typical or standard Dialogue Box in Minitab at present, but let us look at the one for the histogram to illustrate the parts mentioned above. You should get it up on the screen by using the Menu Command:

```
Choose Graph▶ Histogram
```

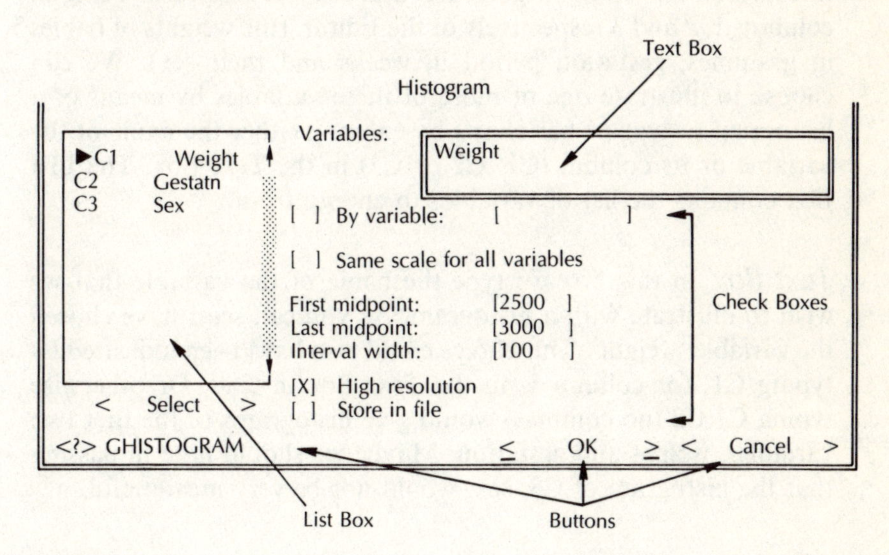

Figure 2.1

The main keys that we use to navigate amongst the elements of the Dialogue Boxes are the Tab and Arrow keys, and we register our choices when we have arrived at the desired destination with either the Enter key or the Space Bar.

After you have brought the Histogram Dialogue Box up on the screen you should compare it with figure 2.1. It should look approximately the same except that the various boxes will be empty, assuming that your Data Editor has no data in it. Move between the elements of the Dialogue Box by pressing the Tab key a few times and noting the movement of the cursor. It starts off in the Text Box and one depression of the Tab key takes it to the next element, marked 'By variable'; after eleven depressions we are back in the Text Box. In every Dialogue Box the cursor will perform similar closed circuits at the command of the Tab key. At each point of rest on such a circuit we can register a choice, should we so wish, by depressing the Enter key, the Space Bar or typing in instructions etc., as appropriate. Let us now look at the various parts of the Histogram Dialogue Box in more detail.

List Box This box contains a list of the variables that are in the columns of the Data Editor. In the List Box shown in figure 2.1, three variables – weight, gestation and sex – are listed as being in columns 1, 2 and 3 respectively of the Editor. (the weights of babies in grammes, gestation period in weeks and their sex). We can choose to illustrate one or more of these variables by means of a histogram (a type of bar chart) by entering either the name of the variable or its column (Cl, C2 or C3) in the Text Box. The List Box contains the list of variables to choose from.

Text Box In this box we type the name of the variable that we wish to illustrate with a histogram. As you can see I have chosen the variable 'weight'. This choice could also have been indicated by typing Cl, for column 1, in the Text Box instead. Or otherwise typing Cl C2 (no commas) would give histograms of the first two variables, weight and gestation. Maybe we should note in passing that the histogram of C3, sex, would not be very meaningful.

Check Box These are various smaller 'boxes' in which we can register various choices, usually optional choices. As you can see some of these 'boxes' do not look like boxes at all. On the diagram you will notice that I have entered 2500 in the Check Box which stipulates the first midpoint of the histogram and 3000 in the last midpoint box. I have also specified an interval between the bars of the histogram of 100. If nothing had been entered in these three Check Boxes then Minitab would have used default values. That is, it would choose what it considered to be appropriate values for the first and last midpoint and interval. These default values will not be registered in the Check Boxes.

It is important to note here that in some Dialogue Boxes the circuit made by the cursor on the repeated pressing of the Tab key will only highlight the first Check Box in a group (field) of Check Boxes. If this is the case it is necessary to bring the cursor to the first Check Box in the group and then move to the one that you want with the Arrows. The Space Bar is then usually used to register your choice.

Buttons The elements to be found at the bottom of the Dialogue Boxes, such as OK, Cancel, <?> etc., are called Buttons and the circuit of the cursor under the control of the Tab key always includes these elements. Once the cursor is positioned on a Button by using the Tab key we register the choice by pressing the Space Bar.

The two most important Buttons are the OK and the Cancel. Positioning the cursor on the Cancel Button and pressing the Space Bar will cancel all the choices made so far in the Dialogue Box. You can use it if you discover a serious mistake in your entries or change your mind and wish to start filling in the Dialogue Box afresh.

The OK Button is chosen when you feel that everything in the Dialogue Box has been filled in to your satisfaction. Pressing the Space Bar, for instance, when the cursor is on the OK Button will exit you from the Histogram Dialogue Box and draw the histogram of the variable 'weight' according to the specifications entered in the various boxes. Choosing the <?> Button will give Help on the subject of the Dialogue Box; for instance in the box that we have been considering we would get Help regarding histograms.

As you can no doubt see, navigation in and through the various Dialogue Boxes is not as straightforward as in the case of the Menu Boxes. You will probably need a reasonable amount of practice in moving in the Dialogue Boxes in order to gain proficiency.

You may find at this stage, along with many newcomers to this system, that you have difficulty in knowing which choices to enter with the Space Bar and which to enter with the Enter key. Sometimes either will do but in other cases we do need to be specific. It is worth remembering that pressing the Enter key in some places may have the same effect as going to the OK Button and choosing it with the Space Bar. That is, you may have the frustrating experience of being shown to the exit (i.e. OK) before you have completed all the Check Boxes that you intended. Don't despair, such minor frustrations are a part of computing. Press the Escape key and start again.

In the case of some of the procedures that I have outlined so far in this chapter there are different ways of achieving the same results, for instance by other commands or procedures or indeed Mouse Commands. However, in order not to confuse the newcomer I have stuck to what I consider to be the most basic ways of navigating amongst and in the boxes. By studying these and practising on the keyboard (and not forgetting the Help menus) you should easily gain the necessary proficiency to work through the exercises in this chapter.

Some Minitab Commands

Before starting the exercises it may be useful if I gather together a few commands, both of the Menu and of the Direct type, and list a few keys. Most of these will be unfamiliar to you as yet, but I gather them together here for later reference. More details can of course be found in the *Minitab Reference Manual* or through the Help menu.

C:\>MINITAB	Start Minitab
MTB > STOP	Exit from Minitab
Alt+X	Exit from Minitab
Alt+D	Primary Screen to Data Editor
Alt+M	Data Editor to Primary Screen
Alt+F	File menu
Alt+E	Edit menu
Alt+C	Calculate menu
Alt+S	Statistics menu
Alt+G	Graph menu
Alt+O	Directly to OK in Dialogue Box
Alt+L	Back to last Dialogue Box

Alt + H	Help menu
<?> or F1	Help menu
MTB > HELP	Help menu
MTB > HELP HISTOGRAM	Help regarding a specified area
Escape key	Escape from Menus
Tab key	Move cursor in closed circuit in Dialogue Box
Space Bar	Enter some choices in Dialogue Boxes
Enter key	Enter Direct Commands and some choices in Boxes etc.
Arrows	Move around Primary and Data Screens, Menu options and elements in a field etc.
F10	Menu when in Data Editor
F6	Big carriage return in Data Editor
F3	Directional arrow change in Data Editor
F8	Delete a cell in the Data Editor
A:	Refers to the floppy disk in drive A
C:	Refers to the hard disk C

MTB > SAVE 'A:\JANE01'	Saves data in Editor to a Minitab file in A: called JANE01.MTW
MTB > RETRIEVE 'A:\JANE01.MTW'	Retrieves the Minitab data file JANE01.MTW from A: into Editor
MTB > RETRIEVE 'C:\MINITAB\DATA\PULSE.MTW'	Retrieves the Minitab data file PULSE.MTW from directory DATA, which is in directory MINITAB in drive C:
MTB > READ 'A:\DATA03.DAT' into C1-C3	Reads ASCII file DATA03.DAT from A: into Editor

MTB >#	Starts a line of comment
MTB >OW=70	Set output width (here 70)
MTB >OUTFILE 'A:\YASMIN'	Puts output of analysis into a file in A: called YASMIN.LIS
MTB >NOOUTFILE	Stops output of analysis
MTB >DIR A:	Displays directory of disk in A:
MTB >DIR C:	Displays the directory of Minitab in C: (i.e. in the hard disk)
MTB >DIR C:\MINITAB\DATA	Displays the directory called Data which is in the directory of Minitab in C:
MTB >TYPE 'A:\GZALA02.DAT'	Displays the contents of file in A: called GZALA02.DAT
SUBC> ABORT	Exits from a failed sub-command

It is also necessary before starting the exercises to make a few general points and say a bit about certain aspects of Direct Commands.

1 Each Direct Command must start on a new line of the Primary Screen. After typing the command it is entered with the Enter key.
2 Minitab Direct Commands, like other typed computer commands, are separated from the argument (the element or elements to which the command applies) by at least one space, e.g. MTB >PLOT C1 against C2. Here a space separates the Direct Command PLOT from the arguments C1 and C2 (the values of the variables in columns 1 and 2).
3 Direct Commands can be typed in after the Minitab prompt, MTB >, in upper- or lower-case letters. However, in this

book, the parts of the commands that I have shown in lower case are optional and not essential to the structure of the command. They are given as an aid to help you to understand what the command does. It is probably wise for you to use such optional parts, at least initially, as they will also help you to make sense of print-outs when you study them later. For example in the command given in point 2 above, the word 'against' is not strictly necessary to the command and MTB >PLOT C1 C2 or MTB >plot c1 c2 would have sufficed.

4 Direct Commands can usually be shortened to four or so letters, usually the first four. The exact contraction can be found by trial and error or by referring to the *Minitab Reference Manual*. e.g. MTB >HISTOGRAM C1 can be shortened to MTB >HIST C1.

5 Values of the variables are entered into columns in the Data Editor, which are numbered C1 (column 1), C2 (column 2) etc. Such columns can also be named, usually in a way which indicates the type of variable that we have in the column. In Direct Commands, and indeed in Menu Commands, we can use either the column number or the column name. It is usually quicker and more convenient, however, to use the column numbers. When we are referring to a succession of columns they can be abbreviated by a hyphen between the first and the last.

For example, if we have various values of the named variables weight, height and age in columns C1, C2 and C3 respectively, we can get histograms of the data by any of the following Direct Commands:

```
MTB > HISTOGRAM C1 C2 C3
MTB > HIST C1-C3
MTB > HIST 'WEIGHT' 'HEIGHT' 'AGE'
```

Note that in the last case we have to enclose the names of the columns in single inverted commas.

6 File names and pathways must also be enclosed in single inverted commas. For example,

```
MTB > RETRIEVE 'C:\MINITAB\DATA\MEATLOAF'
```

The file MEATLOAF.MTW has a pathway which tells the computer that it is to be found in the directory called DATA, which is in the directory called MINITAB which is on the hard disk C.

7 Remember that the number zero, 0, and the letter O, are two different symbols, as are the number one, 1, and the lower case letter l.

8 It is important to read carefully all messages and prompts that appear on the screen, especially the error messages. If you find that you cannot understand a particular error message you should use the Help menu or the *Minitab Reference Manual* to help you.

9 In parts of this chapter you will find it necessary to separate the grammar or punctuation of the text from the Direct Commands. Consider the following sentence:

Type in and enter the Direct Command MTB > SAVE 'A:\NAJMUS'.

A bit of nous tells us that the command ends with the last inverted comma after the file name NAJMUS and that the full-stop is a part of the English sentence. Commands that end in full-stops are rare in computing and the only such commands that we deal with in this chapter are final subcommands (see exercises 3 and 4).

10 Lastly, do not attempt to remove the floppy disk from the disk drive if the disk drive light is on. Apart from spilling coffee or beer on the keyboard, this is virtually the only way you are likely to damage the system inadvertently.

After many pages of preamble we now come to the exercises which form the core of this chapter.

You will find that I have worked through each of the following exercises in two ways, first using the Menu Commands and then again using the corresponding Direct Commands. I then print out an abbreviated version of the Direct Commands entered onto the Primary Screen to serve as a summary for each pair of exercises.

You should remember that it is no good merely reading each exercise, you need to read it through quickly and then follow it step by step on the computer. This is the only way to make this 'book knowledge' of Minitab *your* knowledge.

Also remember what I said earlier about the objectives of the present chapter. These exercises and the whole chapter are *not* an attempt to teach you data analysis. The aim here is more basic and is simply to help you to find your way around the computer package and perhaps to provide you with a few brief models of how one can go about various necessary routine tasks and operations, such as entering data, saving data on disk, recording analysis sessions and so on. As you work through these examples do not worry if you do not fully understand some of the statistical and data analysis terms (e.g. histogram, stem-and-leaf plot, mean, correlation etc.) The specifically data analysis aspects of this book will be covered in subsequent chapters. Here your task is to develop your confidence in the keyboard use of Minitab.

Eight Hands-on Exercises

Exercise 1m (m for Menu Commands)

In this exercise we enter a small data set of the weights etc. of 10 premature babies into the Data Editor and then use various Menu Commands to get a simple analysis or picture of the data. After entering the data in the Editor we will save it onto a floppy disk, print the data into the analysis file, get a statistical description, and finally plot and correlate two of the variables.

You should try to work this exercise through from beginning to end; however, if you do get stuck you should note that it is in sections. If you go adrift you can return to the start of the section (probably by using the Escape key) and work it again. If you still stick, go on to the next section (unless you get stuck on the first!).

The data set is as shown below. The weights of the 10 babies are in grams and the gestation period is given in weeks to the nearest week. In the Sex column, 0 indicates a female and 1 a male.

Weight	*Gestation*	*Sex*
2500	33	1
2550	33	1
2700	34	0
2600	36	1
2700	36	0
2750	37	0
2800	38	0
2900	38	1
2850	39	0
2950	40	1

To start enter Minitab with the appropriate command on your 'shell' menu or use the DOS command C:\>MINITAB.

To enter the data into the Editor Alt+D to go to the Editor. The cursor should now be in the first cell of the first column C1,R1 (if not position it with the Arrow keys); type in the first value of the variable 'weight', 2500, and press Enter. The cursor will now move down to the second cell of the first column, C1,R2; type 2550 and press Enter. If the cursor moves across the screen rather than down, press F3 to ensure that the arrow at the top left-hand corner of the screen is pointing down.) Type and enter all the remaining weights into column 1. Move to the second column with the Arrows or by pressing F6 (big carriage Return) and type in and enter the gestation periods in the appropriate cells; then go to the third column and put in the sex coding for each baby. If you should make a mistake, reposition the cursor in the cell using the Arrow keys and retype the entry.

We have now entered all our data, so with the Arrows move to the top of the Data Screen above C1, type WEIGHT and press the Enter key. Move with the Arrows to above C2 and then C3 and type and enter GESTATN (gestation is one letter too long for a name) and SEX respectively. Alt+M to return to the Primary Screen.

To copy the data to a floppy disk in drive A: First check that you have a formatted floppy in A:

```
Choose File► Save Worksheet As...►
Save As Dialogue Box► Minitab Worksheet►
Select File► TextBox,Filename,type A:\BABIES01► OK
```

To remind you, this means: Alt+F to get the File Menu, then down to Save Worksheet As... with the Arrows and press Enter. In the Save As Dialogue Box choose Minitab Worksheet, Enter. In the new Save As Dialogue Box type A:\BABIES01 into the Text Box called File Name, then use the Tab key to move to OK and press the Space Bar.

We now have a copy of our data file on the disk in drive A: and it is called BABIES01.MTW. The tag MTW is given automatically by the system and indicates a saved Minitab Worksheet, i.e. a data file. This data file that we now have on a floppy disk in A: can be called back into Minitab at a later date. We shall in fact use this stored data in exercise 2.

To get basic information about the data

```
Choose Edit▸ Get Worksheet info...▸ OK
```

This means Alt+E to get the Edit Menu, then down Arrows to Get Worksheet info and Enter. In the Get Worksheet info Dialogue Box, go to OK with the Tab key and press the Space Bar. This gives a brief summary of the data regarding the number of columns and values etc.

To print the Data File onto the Primary Screen

```
Choose Edit▸ Display Data...▸
Text Box,type C1 C2 C3▸ OK
```

This means Alt+E to get the Edit Menu, then down Arrows to Display Data... and Enter. In the Display Data Dialogue Box go to the Text Box and type in C1 C2 C3 (spaces between the columns but no commas). Go to OK with the Tab key and press the Space Bar.

This gives us a copy of the data file on the Primary Screen. The record of what appears on the Primary Screen will constitute our analysis file and it is generally a good habit in computing to print the data file, if it is not large, into the beginning of the analysis file.

To get a statistical description of the first two variables

```
Choose Stat▸ Basic Statistics>>▸
Descriptive Statistics▸ Text Box,type C1 C2▸ OK
```

This means Alt+S to get the Statistics Menu, down Arrows to Basic Statistics > > and Enter. This gives the submenu Descriptive Statistics, Enter. In the Descriptive Statistics Dialogue Box go to the Text Box and type in C1 C2 (spaces but no commas), then to OK with the Tab key and press the Space Bar. This gives a statistical description of the two variables weight and gestation in terms of the number of observations, mean, median and standard deviation etc. The first few values of this description are

```
          N      Mean     Median
Weight    10     2730.0   2725.0 etc.
Gestatn   10     36.400   36.50  etc.
```

To plot Weight against Gestation

```
Choose Graph▸ Scatter Plot...▸ Vertical Axis,C1▸
Horizontal Axis,C2▸ OK
```

This means Alt+G to get the Graph Menu, then down Arrow to Scatter Plot... and Enter. In the Scatter Plot Dialogue Box go to the Vertical Axis Text Box and type in C1, Tab key to Horizontal Axis Box, type in C2, go to OK with the Tab key and press the Space Bar.

This should give you a scatter plot of the two variables weight and gestation. If you look at the graph there does seem to be a pattern to the plotted data in that we could probably fit a straight line roughly across the points from bottom left to upper right. As we shall see later, in chapter 6, this indicates a possible positive correlation between the variables weight and gestation, i.e. the greater the gestation period the greater the weight. Q to quit to the Primary Screen.

As the last part of our first analysis let us calculate a correlation coefficient between the variables that we have plotted.

To find a correlation coefficient for Weight and Height

```
Choose Stat▶ Basic Statistics>>▶ Correlation...▶
Text Box,C1 C2▶ OK
```

This means Alt+S to get the Statistics Menu, down Arrow to Basic Statistics> >, Enter to get the submenu, then down Arrow to Correlation, Enter. In the Correlation Dialogue Box go to the Text Box with the Tab key and type in C1 C2, go to OK with the Tab key and press the Space Bar.

You should find that this gives a correlation coefficient of 0.916 between weight and gestation. As we shall see in chapter 6 this is a fairly high correlation for only 10 pairs of data.

Let us now end this first analysis and take a break by typing MTB >STOP or Alt+X.

Postscript to Exercise 1m We should note in relation to the last section that it is always safer to exit from Minitab using Alt+X rather than the Direct Command STOP, as this gives us an Exit menu. This menu asks us, amongst other things, whether we wish to save the worksheet (i.e. the data in the Editor). We have been virtuous in this present exercise and have saved it already. However, as any researcher will tell you, it is something that is easy to forget. Such a prompt in the Exit menu should prevent us losing the data in the Editor if we have forgotten to save it.

Let us suppose that we had not saved the data; we would use the Exit menu as follows. With the Arrow key go down to Save Worksheet, Enter, and fill in the file name box in the Save As Dialogue Box with the chosen file name, say A:\FIRSTGO, Tab key to OK and press the Space Bar. This saves the data in a file called FIRSTGO.MTW on the disk in drive A:.

We shall now do the same exercise using the corresponding Direct Commands which we will type onto the Primary Screen after the Minitab prompt MTB >.

Exercise 1d (d for Direct Commands)

Enter Minitab with the appropriate command from your 'shell' menu or the DOS command C:\>MINITAB. If you have not in fact left Minitab after Exercise 1m then you will need to clear the Data Editor with the command MTB > ERASE C1 C2 C3. Go to the Editor with Alt + D and enter the data set into columns C1, C2 and C3 and name them as before. Alt + M to return to the Primary Screen. We shall now save the data to a floppy disk in drive A: with the Direct Command,

```
MTB > SAVE 'A:\BABIES02'
```

Note that I have used the file name BABIES02 to differentiate it from the file BABIES01 that you should already have on the floppy disk from exercise 1m. Minitab gives the new data file an MTW (Minitab Worksheet) tag and we now have two copies of our data file on the floppy in A:, i.e. BABIES01.MTW and BABIES02.MTW. (This can easily be checked, if you wish, by asking to see the directory of A: with MTB > DIR A:.)

It is important to note that when we quote a file name in a Direct Command we must enclose the file name and location or pathway (here A:\) in single inverted commas.

Type INFO after the Minitab prompt and Enter to get information about the data. MTB > INFO. This gives:

```
Column   Name      Count
C1       Weight    10
C2       Gestatn   10
C3       Sex       10
```

Type as follows after the Minitab prompt and Enter.

```
MTB > PRINT C1 C2 C3
```

This gives us a copy of the data file in our analysis file as follows:

```
Row   Weight   Gestatn   Sex
1     2500     33        1
2     2550     33        1
3     etc.     etc.      etc.
```

Type as follows after the Minitab prompt to get a statistical description, and Enter.

```
MTB > DESCRIBE C1 C2
```

This gives a description of the first two variables that starts off as follows:

```
          N      Mean      Median
Weight   10    2730.0     2725.0  etc.
Gestatn  10    36.400     36.50   etc.
```

Type as follows to get a plot of one variable against the other: MTB > PLOT C1 against C2. This should give you a scatter plot, but as you will notice it is slightly different to the Menu scatter plot that we obtained in Exercise 1m. (If you wish, an exact copy of this can be obtained by typing as follows: MTB > GPLOT C1 against C2.)

To get the last item of the analysis we type as follows and Enter it. MTB > CORRELATE C1 against C2. As before, this gives a correlation coefficient of 0.916 between the two variables weight and gestation.

Let us end this Direct Command presentation by typing and entering the STOP command: MTB > STOP. This stops the program.

As you can see these Direct Commands are considerably shorter to execute and explain. However, as I mentioned before it is certainly worth learning to use both the Menu and the Direct Command systems, despite the apparent initial complexity of the Menu Command system.

As I also mentioned earlier, most Direct Commands can be shortened and words like 'against' etc. can be left out all together, but remember that they do help us to understand commands, especially in the beginning. I give below a summary of the few Direct Commands used in this exercise after we entered the data on the Editor.

```
MTB > ERASE C1-C3 (if continuing directly from exercise 1m)
MTB > SAVE 'A:\BABIES02'
MTB > INFO
MTB > PRINT C1-C3
MTB > DESC C1 C2
MTB > PLOT C1 against C2
MTB > CORR C1 against C2
MTB > STOP
```

This ends the first pair of exercises. I hope you managed to complete them successfully and are in confident mood to tackle the next pair. In these you will find that I have not given such lengthy instructions, especially in the Menu Command section.

Postscript to Exercises 1m and 1d Whilst bearing in mind that the main thrust of this chapter is to give you the confidence and ability to move around the Minitab terrain, you should note that, in this first pair of exercises, we have omitted to do something that is very important in the everyday practice of data analysis. Although we saved our data file we did not save our analysis.

Prudent researchers will, of course, always save their analyses and keep them safe until they are superseded. In the next short exercises we shall learn how to save our analyses to disk.

Exercise 2m

In this exercise we shall retrieve the data file of babies' weights etc. that we saved in exercise 1. We shall then set up the Outfiling procedure which allows us to record or save our analysis to an external file, and proceed with a simple analysis of the data. We shall rank the babies' weights and then dotplot them according to sex. Then we shall save the modified data file, finish recording our analysis and exit Minitab.

To retrieve the data into the Editor If you completed every part of exercise 1 you should have a surfeit of baby data files on your floppy disk, each containing identical data, i.e. files BABIES01.MTW, BABIES02.MTW and FIRSTGO.MTW. Retrieve one of these, say the first, with the following Menu Commands.

```
ChooseFile▸ Open Worksheet▸ Minitab Worksheet▸
Open File▸ Filename,A:\BABIES01.MTW▸ OK
```

If you find that you cannot remember how to follow the above Menu Command and its navigation instructions, please go back to exercise 1m to refresh your memory.

Before going on to the next section, check that the data are indeed in the Data Editor with Alt + D after you have completed the above command.

To save our analysis into a file (OUTFILE)

```
Choose Files▸ Other Files>>▸
Start Recording Session...▸ Select File▸
Record Session As▸ Filename,A:\BABIES01▸ OK
```

This Menu Command will send all our subsequent analysis to a file in the floppy disk in A: called BABIES01.LIS. Minitab automatically assigns it the tag LIS to indicate its status as an analysis or list file, i.e. it lists our results.

You will find that is a good policy to call the analysis file something similar to your data file so that you end up with a recognizable pair of files, BABIES01.MTW and BABIES01.LIS. This bracketing together of data and analysis files is essential once you start to accumulate a few files. We need to ensure that, at some future date, data files and their corresponding analysis files are recognizable by their names as belonging to a related set of files.

To rank the weights of the babies

```
Choose Calc▸ Rank...▸ Rank Dialogue Box,
Rank Data in,C1▸ Put ranks in,C4▸ OK
```

In the output from this command you should note how the ranks are assigned. The weight in R1 is the least and is given the rank of 1, the weight in R2 is the next lightest and is given the rank of 2, R3 is the next lightest and has a rank of 3, and so on. The heaviest baby is given a rank of 10. Note, however, that where two babies have the same weight they are ranked equally but the rank will not be an integer (whole number). For instance R3 and R5 of C1 are both 2700 g and so they are both given the average of the next two rankings, 4 and 5, which is of course 4.5. We have now 'used up' ranks 4 and 5, so the next lightest baby is ranked 6.

To dotplot weights for male and females (on the same axis)

```
Choose Graph▸ Dotplot...▸ Dotplot Dialogue Box▸
Text Box,variables,C1▸ By variable,X▸
By variable,C3▸ OK
```

(Note: In this command you need to enter an X in the Check Box on the left of the By variable before you can enter the By variable Check Box with the Tab. key, in order to type in C3.) This gives two separate dotplots of the weights on the same axis. Each dot represents the weight of a baby. You should note that, for some reason, the weights of the female babies are grouped nearer to the average weight (mean found earlier to be 2730 g and the median 2725 g) whilst the male babies are in two groups one at each end of the distribution of weights. If our focus in this chapter was on data analysis rather than computing, this would certainly merit investigation.

To save our data to a floppy in drive A

```
Choose File▸ Save Worksheet As...▸
Save As Dialogue Box▸ Minitab Worksheet▸
Select File▸ Text Box,Filename,A:\BABIES03▸ OK
```

This gives us a copy of our amended data file, adding one more baby data file to our collection in A:. This is not of course identical to the others as it contains the weight rankings.

To finish recording our analysis and exit from Minitab

```
Choose File▸ Other Files>>▸
Stop Recording Session...
```

This gives us a copy of our analysis in a file in A: called BABIES01.LIS. Exit Minitab with Alt + X.

We shall now do the same exercise using the corresponding Direct Commands.

Exercise 2d

Enter Minitab with the appropriate command or, if you are still in Minitab after having completed exercise 2m, clear the Data Editor with the command MTB > ERASE C1-C4.

Type and enter the Direct Command, MTB > RETRIEVE 'A:\BABIES01' to bring the data from the file on the floppy to the Data Editor.

To save your analysis to a file type the command, MTB > OUTFILE 'A:\BABY'. This will give you an analysis file in A: called BABY.LIS. You should note that I have chosen a different file name here from that used in exercise 2m.

To rank the weights of the babies we use the command,

```
MTB > RANK C1 and put in C4
```

To dotplot the male and female weights on the same axis the command is,

```
MTB > DOTPLOT C1;
SUBC> BY C3.
```

You should note that this command is in the form of a couplet or small 'poem' (computer haiku?), where the main command DOTPLOT is followed by a subcommand (SUBC). We can have up to four subcommands following the main command. It is important to notice the punctuation: the semicolon at the end of the main command brings up the SUBC > prompt for the first subcommand. After entering the first subcommand, here BY C3, we must end the couplet with a full-stop.

If there had been more than one subcommand, each line of the 'poem' would have ended with a semicolon except the last which must always end with a full-stop. The BY C3 subcommand above uses the sort code in C3 (0 for females, 1 for males) to produce two separate plots on the same axis.

Minitab has many of these composite commands, where a command is followed and qualified by a subcommand. You should note that they usually correspond in the menu system to entering something into a Text Box and then qualifying it by an entry or entries in Check Boxes.

We will now save our modified data file with the command, MTB > SAVE 'A:\BABIES03' which gives us yet another baby data file on the floppy in A:.

To finish recording our analysis you should use the command MTB > NOOUTFILE. Note that this command is not two words but one group of letters with no spaces. All being well, we should now have recorded all the above analysis in a file in A: called BABY.LIS.

You should now end and exit Minitab with MTB > STOP.

Below, I have given a summary of the Direct Minitab Commands that we have used in this exercise.

```
MTB > ERASE C1-C4  (if you continued directly from exercise 1)
MTB > RETRIEVE 'A:\BABIES01'
MTB > OUTFILE 'A:\BABY'
MTB > RANK C1 and put in C4
MTB > DOTPLOT C1;
SUBC> BY C3.
MTB > SAVE 'A:\BABIES03'
MTB > NOOUTFILE
MTB > STOP
```

Postscript to Exercises 1m, 1d, 2m and 2d In these first two pairs of exercises we have probably covered enough of the Minitab basics for you to appreciate the computing element in subsequent chapters, relying of course on the Help programme as necessary. However, Minitab is a large package and in the following exercises I will introduce you to a few further groups of commands but my main intention in these remaining exercises is to help you to consolidate your existing knowledge of Minitab and so give you greater confidence to explore on your own. Please complete the remaining exercises if you have time; it will probably pay dividends in subsequent chapters where you will be able to take the computing element a little more for granted and concentrate on data analysis which, of course, is the prime focus of the other chapters in this book.

Exercise 3m

In this exercise we use the data stored in a file called DATA03.DAT. This is a file on the floppy disk mentioned in chapter 1, which can be obtained from the ESRC at the University of Essex. A print-out of this file is given in appendix 1, and you should look at it before attempting the exercise. Note that it has three columns and it details the sex, heights and weights of a small group of 84 students. We shall enter the data in the Editor and then start recording our analysis. The simple analysis will consist of making boxplots and histograms of the data relating to height. We will save our analysis onto a floppy disk.

Before starting the exercise remember to enter the disk containing the file DATA03.DAT into the disk drive.

To read the data file DATA03.DAT into the Editor

```
Choose File▶ Import ASCII data...▶
Import ASCII data Dialogue Box▶
Text Box,Put data in columns,C1 C2 C3▶ OK▶
Import Text from file, filename,A:\DATA03.DAT▶ OK
```

This command reads the ASCII file DATA03.DAT into the Editor. The letters ASCII (pronounced askey) stand for American Standard Code for Information Interchange. An ASCII file is one which conforms to certain rigid standards regarding format and contents etc. It is a universal standard for computers and it can be used to transfer (interchange) material from one computer package to another. As we shall see later in this exercise, data from the Minitab Editor can also be saved in an ASCII form rather than in the Minitab form (with the tag MTW) used previously, using the Export ASCII data command.

You should note that when importing an ASCII data file we need to specify the columns that it will occupy, in our case C1, C2 and C3.

To save our analysis in a file

```
Choose Files▶ Other Files>>▶
Start Recording Session...▶ Select File▶
Record Session As▶ File name,A:\DATA03▶ OK
```

This will give us a copy of our analysis in a file in A: called DATA03.LIS. The LIS tag is added by Minitab to indicate an analysis or list file.

To Boxplot male and female heights on the same axis

```
Choose Graph▸ Boxplot...▸ Boxplot Dialogue Box,
variable,C2▸ By variable,X▸ By variable,C1▸ OK
```

(Note: In this command you need to enter and X in the Check Box to the left of By variable before you enter Cl in the By variable Check Box use the Tab. key.) This command will give you a boxplot of each distribution of heights, males and females, on the same axis. As we shall see later, in chapter 4, the boxplot is a useful way of showing and comparing the spread of data. The box in the diagram indicates the middle half of the distribution and the whiskers the extent of most of the remaining distribution. Press Q to quit graphics.

To obtain histograms of male and female heights

```
Choose Graph▸ Histogram...▸
Histogram Dialogue Box▸ Text Box variables,C2▸
By variable,X▸ By variable,C1▸
Same scale for all,X▸ OK
```

(Here again you need to enter and X to the left of By variable before you can enter Cl in the By variable Check Box use the Tab. key. Also, to get both histograms on the same scale an X has to be entered to the left of Same scale for all.) You should note that the histogram of the distribution of the heights of the 45 male students appears to be fairly symmetrical and could probably be modelled or described by a Gaussian or normal distribution. We shall look at such matters in chapter 9. Press N for next and then Q to quit graphics.

To finish recording our analysis and exit Minitab

```
Choose File▸ Other Files>>▸ Stop Recording
Session...
```

This command will give you a copy of the analysis in a file in the floppy in A: called DATA03.LIS. We now exit from Minitab with the command Alt + X.

Let us now do the same exercise using Direct Commands.

Exercise 3d

Enter Minitab with the appropriate command, e.g. C:\ > MINITAB, or if you are still in Minitab after having completed the previous exercise, clear the Data Editor with MTB > ERASE C1-C3.

Type and enter the Direct Command, MTB > READ 'A:\DATA03.DAT' into C1-C3, to bring the data from the file in the floppy to the Data Editor. Notice that with this command for an ASCII file we need to specify the columns to read the data into. You should compare this with the RETRIEVE command that we used for Minitab data files.

To save the analysis to file we need to give the command MTB > OUTFILE 'A:\DATA3D', which will give a copy of the analysis in a file in A: called DATA3D.LIS.

To obtain the boxplots of the heights by sex in C1 the command is

```
MTB > BOXPLOT C2;
SUBC> BY C1.
```

You should note the form of this couplet, especially the punctuation.

Next we obtain the histograms with a similarly structured two-line command:

```
MTB > GHISTOGRAM C2;
SUBC> BY C1.
```

or

```
MTB > HISTOGRAM C2;
SUBC> BY C1.
```

To finish recording our analysis we give the command MTB > NOOUTFILE (one word) and then exit with MTB > STOP.

Below I give a summary of the Direct Commands that we have used in this exercise.

```
MTB > ERASE C1-C3 (if you continued directly from exercise 3m)
MTB > READ 'A:\DATA03.DAT' into C1-C3
MTB > OUTFILE 'A:\DATA3D'
MTB > BOXPLOT C2;
SUBC> BY C1.
MTB > GHISTOGRAM C2;
SUBC> BY C1.
MTB > NOOUTFILE
MTB > STOP
```

So far, in these hands-on exercises, we have used small data files with few rows and columns in order that we can more readily see what is going on. These small files are useful, and probably necessary, when we are initially learning data analysis but we need to appreciate that most data files that we shall have to deal with will be much larger.

In this last exercise we look briefly at one of the moderately sized data sets that is built into Minitab. Nearly 30 such data sets are available within Minitab. There is unfortunately no Help programme for these data sets, but details can be found in the *Minitab Handbook* by Ryan, Joiner and Ryan. The data sets can be retrieved with the command
MTB > RETRIEVE 'C:\MINITAB\DATA\file name'.

Some file names are PULSE, LAKE, MAPLE, MEATLOAF, POTATO, PERU, TWAIN. I have chosen the PULSE data file for this last exercise. The directory where the data files are kept can be examined with the command MTB > DIR C:\MINITAB\DATA.

The PULSE data which we will use were collected from 92 students. The students first took their own pulse rate and then tossed a coin. The tails sat down and the heads ran on the spot for one minute; then all measured their pulse rate again. Other information was also collected from them regarding smoking, height, weight, and physical activity rate and was tabulated as follows:

	Variable	*Description*
C1	Pulse 1	First pulse rate
C2	Pulse 2	Second pulse rate
C3	Ran	Ran = 1; did not run = 2
C4	Smokes	Regularly = 1; not regularly = 2
C5	Sex	Male = 1; female = 2
C6	Height	In inches
C7	Weight	In pounds
C8	Physical activity	Slight = 1; moderate = 2; lots = 3

Exercise 4m

In this last pair of exercises we use the PULSE data set detailed above. First we retrieve the data into the Editor and then start Outfiling our analysis. We shall call for information on the data and then print the data into the analysis file. As mentioned earlier in this chapter, this is standard practice among researchers when the data set is not too large; however, in order to save space and time I have only done this once in the previous exercises. We shall then stem and leaf and dotplot some of the data and finally get Minitab to draw up some tables.

To retrieve the data into the Editor

```
Choose File► Open Worksheet...► Minitab Worksheet►
Open File,file name,C:\MINITAB\DATA\PULSE► OK
```

Have a look at the data with Alt+D.

To save our analysis in a file

```
Choose File► Other Files>>► Start Recording
Session...► Select File►
Record Session As,file name,A:\PULSE► OK
```

To get basic information about the data

```
Choose Edit► Get Worksheet Info...► OK
```

To print the data on to the Primary Screen

```
Choose Edit▸ Display Data...▸ Display Data,
Text Box,C1-C8▸ OK
```

To stem and leaf the first and second pulse rates

```
Choose Graph▸ Stem-and-Leaf▸ Stem-and-Leaf
Dialogue Box,variables,C1 C2▸ OK
```

To dotplot the first and second pulse rates on the same axis

```
Choose Graph▸ Dotplot...▸ Dotplot Dialogue Box,
variables,C1 C2▸ Check Boxes,First midpoint,40▸
Last midpoint,140▸ OK
```

You should note that we have plotted the two dotplots on the same axis by specifying the first and last points by means of the Check Box entries. It would be interesting, if we had time, to split the pulse rate data by the sort codes in C3, i.e. according to whether they ran or not. A dotplot of the pulse rate of the students who ran would no doubt show a clear shift, which is partially masked in our present plot by the data for those who did not run.

To tabulate the smoking variable against sex

```
Choose Stat▸ Tables>>▸ Cross Tabulation...▸
Cross Tab. Dialogue Box,Classification variables,
C5 C4▸ Check Boxes,CountsX▸ Column percentsX▸ OK
```

The core of this command draws up a 2×2 table of the 92 men and women by how much they smoke. However, we have used the Check Boxes (Tab to the first, then Arrows down to our choice and enter it with the Space Bar) to also put in the column percentages as well. Look carefully at the table and see if you can work out what the entries mean, e.g. of the 92 students, 28 smoke regularly and of these 20 (71.43%) are men and 8 (28.51%) are women.

As you probably noticed in going through the Check Boxes, we could also have used a chi-square test on the data in the table. We shall look at this very useful test in chapter 11.

You will find as you use Minitab that it is very good at tabulating data. Maybe this is why the researchers at the University of Pennsylvania gave the package its rather quaint name!

To finish recording our analysis and then exit Minitab

```
Choose File▸ Other Files>>▸ Stop Recording Session
```

This gives us a copy of our analysis in a file in A: called PULSE.LIS. Now you should exit with Alt + X.

We shall now do the same exercise using Direct Commands.

Exercise 4d

Enter Minitab or if you are still in it after the previous exercise clear the Data Editor with the command MTB > ERASE C1-C8. Type and enter the Direct Command MTB > RETRIEVE 'C:\MINITAB\DATA\PULSE' to retrieve the data and enter it into the Editor.

To save our analysis to file you should now give the command MTB > OUTFILE 'A:\PULSE', which will give us a copy of the analysis in a file in A: called PULSE.LIS.

To obtain basic information on the worksheet in the Editor the command is MTB > INFO, and as we have decided to print the data into our analysis file we give the further command MTB > PRINT C1-C8.

To display the distribution of pulse rates as stem-and-leaf diagrams we use MTB > STEM-AND-LEAF C1 C2.

As we wish to dotplot the two given pulse rates on the same axis we need to qualify the dotplot command with a subcommand as follows:

```
MTB > DOTPLOT C1 C2;
SUBC> START at 40 end at 140.
```

We can tabulate the variable of smoking by sex with the command,

```
MTB > TABLE C5 by C4;
SUBC> COUNTS;
SUBC> COLPERCENTS.
```

This will give us a table with cells which detail both the number of people and the percentages.

We shall now finish recording our analysis with the command MTB >NOOUTFILE and then exit Minitab with MTB >STOP.

Below I give a summary of the Direct Commands that we have used above.

```
MTB > ERASE C1-C8
MTB > RETRIEVE 'C:\MINITAB\DATA\PULSE'
MTB > OUTFILE 'A:\PULSE'
MTB > INFO
MTB > PRINT C1-C8
MTB > STEM AND LEAF C1 C2
MTB > DOTPLOT C1 C2;
SUBC> START at 40 end at 140.
MTB > TABLE C5 by C4;
SUBC> COUNTS;
SUBC> COLPERCENTS.
MTB > NOOUTFILE
MTB > STOP
```

Printing You will have noticed that in the course of these exercises, although we have saved both the data and analysis files to floppy disk (we could have saved them to the hard disk C: had we wished) we have not printed any of them. In data analysis it is of course necessary to print files and get paper copies which can be studied and annotated and there are two ways to do this: through DOS or with a word-processing package such as Wordperfect or Wordstar etc.

It should be noted, however, that the files that we would like copies of are of two kinds: ones with MTW tags and those with either DAT or LIS tags. The former are Minitab worksheet files (MTW) and cannot be printed out directly using either DOS or a word-processing package. They are Minitab specific and can only be retrieved into Minitab.

We can of course print them indirectly as we saw in the exercises, either by first copying them into an analysis file (as in exercises 1 and 4) and printing that or by retrieving them into the Data Editor and exporting them as ASCII files (as in exercise 2) which can be printed.

Files with LIS (analysis) or DAT tags can be printed directly from the floppy disk through DOS with the DOS command

C:\>PRINT A:\FILENAME, or with a word-processor in the normal way.

You should note, however, that, depending on the settings in you word-processing package or printer, you may need to specify the width of the output from Minitab. This is in order to prevent any wraparound of graphs and tables where the output is too wide. The Menu Command to specify output is

```
Choose Files▶ Other Files▶ Start Recording
Session...
```

In the Start Recording Session Dialogue Box you could then set the width of the output to, say, 70 characters. The Direct Command MTW>OW=70, typed before the OUTFILE command, would also achieve the same result.

Postscript

If you have completed all these exercises you have done well and in subsequent chapters you will be able to concentrate more fully on the concepts that lie behind the analyses that we have encountered in this chapter. This is not, of course, to say that you have learnt everything there is to know about Minitab! However, you should by now have enough hands-on experience and confidence to tackle the computing component of the exercises in the following chapters and also to explore other areas of this extensive computer package.

Should you feel the need for further practice using the Minitab system you could use some of the small data files on the disk mentioned in chapter 1 (they can be examined in appendix 1). Or you may wish to use some of the larger data files in the Minitab Data directory (C:\MINITAB\DATA) which you can list from Minitab with the command MTB > DIR C:\MINITAB\DATA. Any one of these files can be retrieved in a similar way to the PULSE file that we used in exercise 4.

It is also worth noting that the *Minitab Reference Manual*, Release 8, PC Version, November 1991, contains a long tutorial called Quickstart. It gives three tutorial lessons, one relatively simple and two complex.

In the next chapter we shall start our study of data analysis proper.

3 Looking at Data

In this chapter we start with two simple data sets of the heights of 100 males and 100 females and look at various useful ways to organize and present the data so that we can best detect any pattern in it. The idea of an average is introduced and then we look at classes, histograms, frequency polygons and finally the very useful technique of patterning raw data into stem-and-leaf diagrams.

When we look at a set of data, such as the heights of 200 people in table 3.1, what catches our eye first is maybe its volume or size or even the way the data are laid out. These are surface or superficial features, however, and what we need to develop is the ability to look both at and into data. Our task here is to develop this skill or ability to look into data. Data such as these heights are usually referred to as raw or crude data and in this section we shall begin to study a central element in data analysis, i.e. organizing the data and looking for some pattern in it.

Table 3.1 Heights (cm) of 100 males and 100 females measured to the nearest centimetre

Men	173	181	169	183	179	167	183	172	182	175
	176	173	182	173	170	173	172	176	180	177
	186	168	158	175	174	177	170	179	192	164
	175	179	177	177	173	162	160	169	167	175
	182	166	179	174	165	176	174	175	178	174
	179	172	193	171	186	180	173	170	176	181
	172	170	174	171	171	157	174	170	180	162
	181	172	169	175	178	163	172	181	188	188
	175	173	169	166	174	173	174	165	166	171
	177	184	168	169	166	177	168	175	181	175
Women	155	169	162	166	176	162	160	144	159	159
	162	149	167	162	161	163	153	163	153	143
	164	156	159	166	162	158	150	166	160	169
	168	148	170	168	172	156	160	161	157	158
	158	165	170	163	165	162	156	155	157	164
	167	164	154	155	161	154	157	160	162	169
	163	165	158	164	169	161	163	166	153	170
	156	163	161	156	150	171	160	151	163	167
	165	174	170	159	167	165	168	150	172	157
	165	160	154	164	158	160	170	160	164	159

If the original form of the data is taken to be raw, what we are going to do is cook it! I do not of course mean this in a culinary or even a pejorative sense, but the metaphor is apt. We will work with the raw data, prepare them and transform them into a distribution. The word distribution here implies that the data are distributed or patterned in some distinctive way and our task is to bring this pattern to the surface and study it. As suggested, this preparation and presentation as distribution has its parallels in what cooks do with food and, like them, we must see that essential features of the food are not lost in preparation. Crunching data and forcing them into an inappropriate mould will give little satisfaction or insight.

We shall work on the data initially in various and relatively gentle ways which will transform them and make them more meaningful by bringing out any hidden or immanent pattern. A large part of mathematics is concerned with recognizing inherent patterns or regularities and closely related to this is the idea we shall look at in later sections, fitting a chosen model to data.

Some Ways of Organizing Data

To start, however, let us look at the above height data and try to organize them so that we can find or give them some pattern or shape. There are of course a variety of ways of doing this and we shall begin by looking at four:

1 finding the average;
2 grouping the data into classes;
3 illustrating them by histogram and frequency polygon and
4 illustrating them by stem-and-leaf diagrams.

The average

When you first see the two sets of raw data and try to look 'into' them you may think of finding the average value of each of the two sets. This is a useful start and gives us what is called a statistic for each distribution, two numbers instead of 200. These numbers can be thought of as representing the two distributions.

As we shall see later, the word average in mathematics is a generic term covering many types of manipulation but here we can take it to be the common or garden average, the arithmetic mean.

This is most easily found by adding up all the heights in each data set and dividing by 100, the number of people in each set.

$$\text{Average of men's heights} = \frac{173 + 181 + \ldots + 175}{100}\,\text{cm}$$

$$= 174.1 \text{ cm}$$

$$\text{Average of women's heights} = \frac{155 + 169 + \ldots + 159}{100}\,\text{cm}$$

$$= 161.2 \text{ cm}$$

Don't waste time doing these tedious calculations. Take my word that they are correct! In this simple way we have reduced the two sets of data to two figures. We have a statistic for each and as we can see a statistic is one figure which can stand for many.

As we shall see later there are many ways we can reduce data to give us statistics but this statistic, the average, is a start and tells us that the 'general' height level in each set is quite different. The two averages differ by nearly 13 cm (about 5 inches) and the men are in general taller. However, we should note that we have gained no notion of the shape of the distributions hidden in the raw data.

Grouping the data into classes

It is often convenient to organize the set of raw data into subsets. These subsets are most frequently of equal size and called classes. The size of the classes is usually called the class interval and it is chosen by looking at the range of the values of the variable and then deciding on a suitable class interval.

The variable we are dealing with is, of course, height in centimetres and the range of the variable is the difference between the smallest and largest value in each set. Examining the raw data set for men we find the shortest man to be 157 cm and the tallest 193 cm. For the women the corresponding figures are 143 and 176 cm. Therefore the range of the variable in each data set is 193 – 157 = 36 cm (men) and, 176 – 143 = 33 cm (women).

We should notice that although the averages are different there does not appear to be much difference in the spread of the data in each set as measured by the range. As we shall see in a later section the spread of data can be indicated by more reliable or robust statistics, the standard deviation and the midspread.

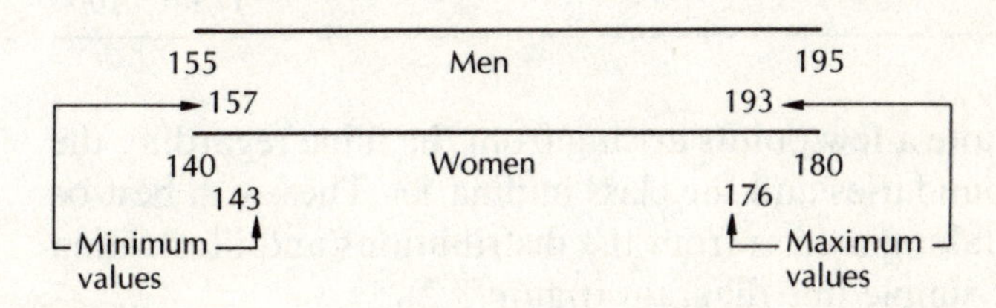

Figure 3.1

The range of each set can be illustrated as in figure 3.1. I have decided to gather the data into classes with a class interval or width of 5 cm, starting with a lower class limit of 155 cm in the case of the men and 140 cm for the women. The upper class limits for men and women will be taken to be 195 cm and 180 cm respectively.

The next task is to count or tally the raw data in each set into the chosen classes as in table 3.2.

Table 3.2 Tally and frequency of height in each class

Class (cm)	*Tally*	*Class midmark*	*Frequency*
Men			
155–159	//	157	2
160–164	/////	162	5
165–169	///// ///// ///// /	167	16
170–174	///// ///// ///// ///// ///// ///// /	172	31
175–179	///// ///// ///// ///// ///// /	177	26
180–184	///// ///// ////	182	14
185–189	////	187	4
190–194	//	192	2
		Total	100
Women			
140–144	//	142	2
145–149	//	147	2
150–154	///// /////	152	10
155–159	///// ///// ///// ///// //	157	22
160–164	///// ///// ///// ///// ///// ///// ///	162	33
165–169	///// ///// ///// ///// /	167	21
170–174	///// ////	172	9
175–179	/	177	1
		Total	100

We should note a few points arising from the table regarding the actual class boundaries and the class midmarks. These can best be dealt with by taking a class from the distributions and illustrating aspects with a simple line diagram (figure 3.2).

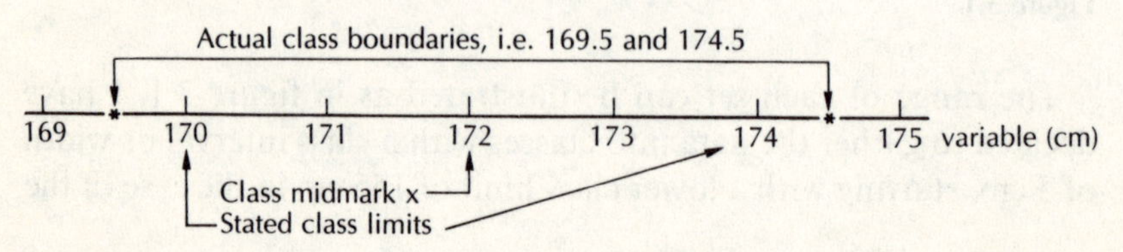

Figure 3.2

The class I have chosen to illustrate is the modal class, 170–174 cm, in the men's distribution of heights. It is the modal class because it has the most heights in it, i.e. it has the greatest frequency, 33. What we need to notice is that although the stated class limits are 170 and 174 cm the actual class boundaries are 169.5 and 174.5 cm. This is a consequence of the original data's being measured to the nearest centimetre. This means that all heights slightly less than 174.5 but greater than 174 cm are included in the class called 170–174 cm. Each class has corresponding class boundaries.

We should also note that the midmark 172 cm is the middle value. Later we shall use the midmark to represent all the data in the class. That is, we shall assume that the modal class data and indeed all class data are distributed evenly or in a balanced way about the class midmark. We should note that this assumption of symmetry or balance of the data in each class about the midmark of each class damages the integrity of the data, but the effect is generally small.

The ideas of actual class boundaries and midmarks that we have looked at above are generally straightforward and usually cause no concern. However, care needs to be taken in ascertaining actual class boundaries and midmarks in some cases, especially when the variable is age.

Let us leave the height data for a moment and look briefly at some classified age data. The age distribution of 40 children admitted in one day to a certain clinic was as shown in figure 3.3.

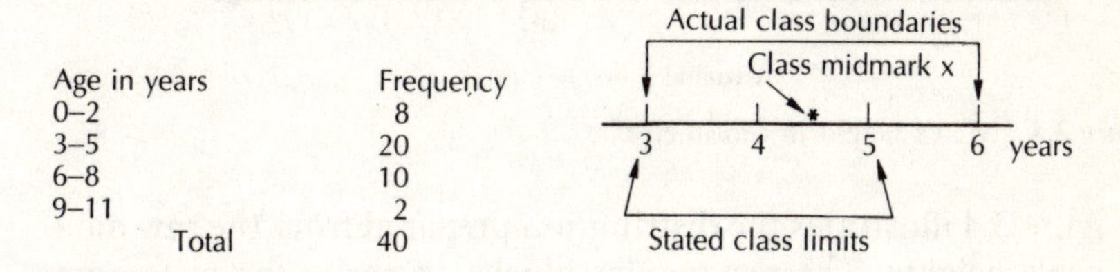

Age in years	Frequency
0–2	8
3–5	20
6–8	10
9–11	2
Total	40

Figure 3.3

Looking again at the modal class, in this case 3–5 years, with the aid of the diagram we see that the actual class boundaries are 3.00 and 6.00 years and the midmark is 4.5 years and not 4 years as may initially be supposed. The boundaries and midmark arise from the way we measure the variable age. For instance a child is classified as 2 years old any time after its second birthday to immediately before its third birthday. The appropriate midmarks for the above classes are 1.5, 4.5, 7.5 and 10.5 years.

The class boundaries and midmarks become important when we seek to illustrate distributions using histograms or frequency polygons. If you are in doubt about such boundaries and midmarks it is advisable to draw a simple line diagram as above.

Let us now return to the male height distribution and illustrate it by a histogram and frequency polygon.

Histograms

A histogram for grouped data is a column graph in which the areas of the rectangular columns are proportional to the frequencies.

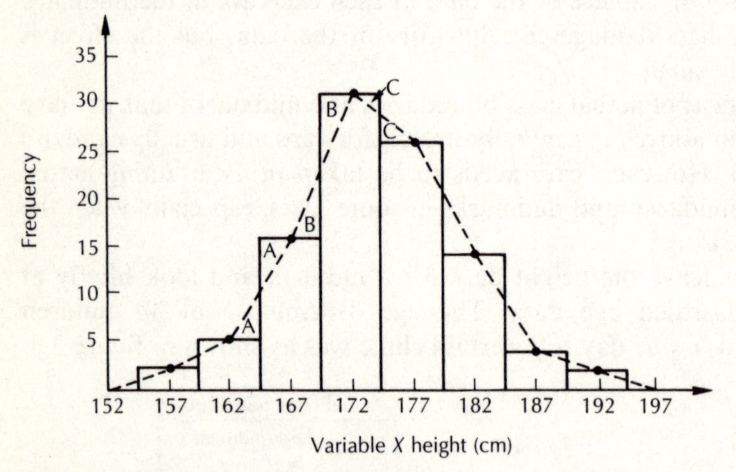

Figure 3.4 Men's height in centimetres

Figure 3.4 illustrates the distribution prepared from the raw data of men's heights. The rectangular blocks comprise the histogram and the frequency polygon is indicated by the broken lines. Let us look at the histogram first.

The width of each rectangular column represents 5 cm, which is of course the class interval. Each column is centred on its midmark and runs between the actual class boundaries. Here all class intervals are of equal width and so the height of each vertical column is proportional to the class frequency. If we had made the class intervals unequal, say at the ends or tails of the distribution, we would have needed to adjust the height of the columns accordingly so that the area continues to represent the frequency. For example if the two end class intervals are made twice as wide as the rest in the distribution then their height must be reduced by half. In this way the area of the columns remains proportional to the frequency.

Histograms are useful and easy to construct especially if we confine ourselves to the most usual type where all class intervals are equal. They can give us a reasonable first picture of the shape or form of the distribution we have formed from the raw data.

Frequency polygons A frequency polygon gives us another picture of the distribution and it is easily obtained from a histogram by joining the middle of the tops of the vertical columns with straight lines. That is, the various class midmarks are plotted against their respective frequencies and the points are joined by straight lines. It is usual to extend the frequency polygon to the horizontal axis. To do this, draw it through the half-height of the end columns and so to the next midmark on the axis outside the distribution (152 and 197 cm in the above example).

It is worth noting, and can be seen clearly from the figure, that the area of the histogram and the area under the frequency polygon and above the horizontal axis are equal. That is, the areas of the corresponding congruent triangles such as A, B and C are equal. We can see that, just as the area of each column of the histogram is representative of the respective frequency, so is the total area of the histogram or frequency polygon representative of the total distribution. We shall look later at the significance of the area under frequency polygons and other curves, i.e. the area bounded by the 'curve' and the horizontal (X) axis. Let us now look at another useful way of preparing the raw data.

Stem-and-leaf diagrams

Stem-and-leaf diagrams are similar in some respects to histograms in that they group the raw data but they are somewhat easier to make and lose no data in a class tallying process.

Each element of the raw data, in our case each height, is split into two parts: a stem and a leaf. These are then ordered to give a distribution. Let us look at the first row of the raw data for men's heights, i.e.

173 181 169 183 179 167 183 172 182 175

Each of these elements may be split into a stem of tens and a leaf of units to give:

17-3 18-1 16-9 18-3 17-9 16-7 18-3 17-2 18-2 17-5

We can make a stem and leaf distribution as follows:

Stem	Leaf	
16	97	
17	3925	
18	1332	
Stem: tens	Leaf: units	N = 10

In the above diagram the stem of 16 and the leaves of 9 and 7 represent the two heights of 169 and 167 cm. There are four leaves and so four heights associated with the stem of 17, i.e. 173, 179, 172 and 175 cm. Similarly the stem of 18 and the associated four leaves represent the heights 181, 183, 183 and 182 cm.

In the completed stem-and-leaf diagram both the stem and the leaves are ordered down and across the page from left to right to give:

Stem	Leaf	
16	79	
17	2359	
18	1233	
Stem: tens	Leaf: units	N = 10

The stem-and-leaf distribution prepared from all (N = 100) of the raw data of men's heights is as follows:

Stem	Leaf
15	78
16	0223455666677888899999
17	000001111222222233333333444444445555555556666777777889999
18	000 111112223346688
19	23

Stem: tens Leaf: units N = 100

The pattern of the distribution does begin to emerge here but we would get a better presentation if we split each of the stems into, say, two parts. The first part will carry the leaves 0 1 2 3 4 and the second the leaves 5 6 7 8 9. We can decide on the way we split (by 2, 5 or 10) or even increase the size of the stem. A fivefold split of the stem would carry leaves of 01 23 45 67 89. For our height data, however, a twofold split of the stem seems suitable and will give groupings similar to the histogram we constructed earlier. We should note, however, that stem-and-leaf diagrams do not lose data as does histogram classification.

The men's distribution of heights now appears as follows:

Stem	Leaf
15	
15	78
16	02234
16	55666677888899999
17	00000111122222223333333344444444
17	5555555556666777777889999
18	00011111222334
18	6688
19	23
19	

Stem: tens Leaf: units N = 100

Back-to-back stem-and-leaf plots The raw data from the set of women's heights can of course be prepared in a similar fashion to men's to give

Stem	Leaf
14	34
14	89
15	0001333444
15	5556666677778888899999
16	000000001111122222223333333444444
16	555555666677778889999
17	000001224
17	6

Stem: tens Leaf: units N = 100

If we now wish to make an initial comparison of the pattern of the two distributions we can plot them back to back.

Men's heights Leaf	Stem	Leaf Women's heights
	14	34
	14	89
	15	0001333444
87	15	5556666677778888899999
43220	16	000000001111122222223333333444444
9999988877666655	16	555555666677778889999
4444444433333333222222111100000	17	000001224
99999887777776666555555555	17	6
43322211111000	18	
8866	18	
32	19	
	19	

N = 100 Stem: tens Leaf: units N = 100

The stems are ordered from top to bottom as before but the leaves for the both distributions are ordered outwards from the stem.

Postscript

In this chapter we have looked at some simple ways to organize raw data and bring any underlying patterns to the surface. A distribution of data can be ordered into classes and then plotted as a histogram or frequency polygon. We also examined the stem-and-leaf technique, which we will also find useful in the next chapter when we find the median and quartiles of a distribution.

We also briefly introduced the idea of an average, and in the next chapter we turn our attention to averages and various other numerical ways to describe or summarize data.

Exercises

1 (i) Using a class interval of 10 cm and starting with a class limit of 150 cm construct a table of grouped data for the first 50 men's heights (i.e. use only the top five rows) in the table at the start of the chapter.

(ii) Draw a line diagram of the first class, 150–159 cm in the above distribution indicating the class limits, actual class boundaries and the midmark.

(iii) What are the midmarks of the other classes?

(iv) Sketch a histogram for these grouped data.

2 The weights of 100 people in kilograms measured to the nearest kilogram are distributed as follows:

Weight (kg)	*frequency*
60–62	10
63–65	22
66–68	26
69–71	20
72–74	18
75–77	4
Total	100

(i) Which is the most popular or modal class?
(ii) Draw a line diagram for the modal class and indicate the class limits, the actual class boundaries and the class midmark.
(iii) What is the class interval?
(iv) Sketch a histogram and frequency polygon of the distribution.

3 The stem-and-leaf distribution below illustrates the women's heights data from the table at the beginning of the chapter.
(i) What makes it different from the women's height distribution shown in the back-to-back diagram?
(ii) Which distribution is to be preferred and why?
(iii) What do you think the column of figures on the left of the distribution indicates?

1	14	3
2	14	4
2	14	
4	14	89
8	15	0001
11	15	333
17	15	444555
26	15	666667777
36	15	8888899999
49	16	0000000011111
(14)	16	22222223333333
37	16	444444555555
25	16	66667777
17	16	8889999
10	17	000001
4	17	22
2	17	4
1	17	6

Stem: tens Leaf: units N = 100

4 Before attempting this question you should make sure that you have done at least some of the 'hands-on' Minitab exercises in chapter 2. The two data sets of men's and women's heights are in the file DATA02.DAT on the floppy disk mentioned in chapter 1. Have a look at the form of the data in this file as shown in appendix 1.

Read this ASCII data file into Minitab with the command

```
MTB > READ 'A:\DATA02.DAT' into C1-C3
```

Note that the men's data are in C2, the women's in C3 while C1 is occupied by the row numbers.

It is probably best to make a copy of the computing you do on the exercises below so use the OUTFILE and NOOUTFILE commands at the beginning and end of the analysis. For example the OUTFILE command MTB >OUTFILE 'A:\HEIGHTS' will give you an analysis file in A: called HEIGHTS.LIS which will contain all the work you do between the initial OUTFILE and final NOOUTFILE commands.

(i) Print the data sets into the analysis file using the command MTB > PRINT C2 C3 and then obtain a statistical description of the data using the command MTB >DESCRIBE C2 and C3.
(ii) Use the command MTB > HISTOGRAM C2 and C3 to draw histograms of the sets. Look carefully at their form and see how they compare with the histograms we constructed previously by 'hand'.
(iii) Form the two sets into Stem-and-leaf distributions with the command MTB > STEM-AND-LEAF C2 and C3 and again compare them with the distributions constructed earlier.
(iv) Investigate, using the HELP facility or a manual, the histograms and stem-and-leaf functions. What subcommands control class width, starting values etc.? Look at the way these subcommands are chosen in the various Dialogue Boxes.

5 Clear the Data Editor after the last exercise with the command MTB > ERASE C1-C3 and read in the ASCII data file DATA11.DAT with the command MTB > READ 'A:\DATA11.DAT' into C1-C12. A hard copy of this file is to be found in appendix 1 and you should study it for a few minutes and note what the various columns refer to. In particular note that column C7 contains the infant mortality rate, IMR, for the listed countries.

Outfile your analysis with the command MTB > OUTFILE 'A:\IMR'. Name the column C7 in the Editor by positioning the cursor above the column with the Arrow keys, typing in the initials IMR and then entering.

(i) Draw a histogram of the IMR data with the command MTB > HISTOGRAM C7.
(ii) Draw a stem-and-leaf diagram of the IMR data with the command MTB > STEM-AND-LEAF C7.
(iii) Obtain a statistical description of the IMR data with the command MTB > DESCRIBE C7.
(iv) Using the above and by referring to the data file, write a few sentences to describe the distribution of the IMR for these 59 countries.

Answers to Exercises

1 (i)

Class cms	frequency f
150–159	1
160-169	10
170–179	30
180–189	8
190–199	1
total	50

(ii)

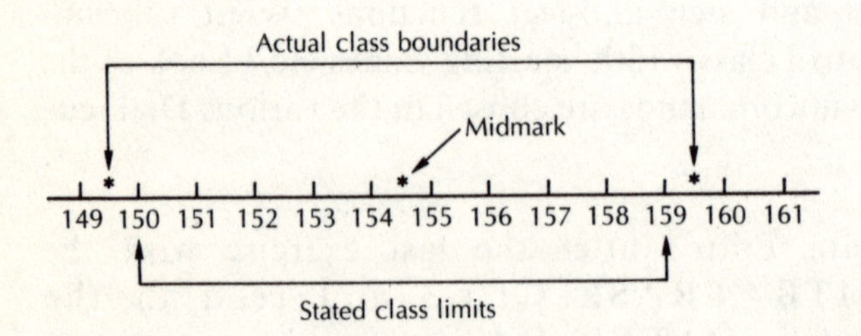

(iii) 164.5, 174.5, 184.5, 194.5 cm

(iv)

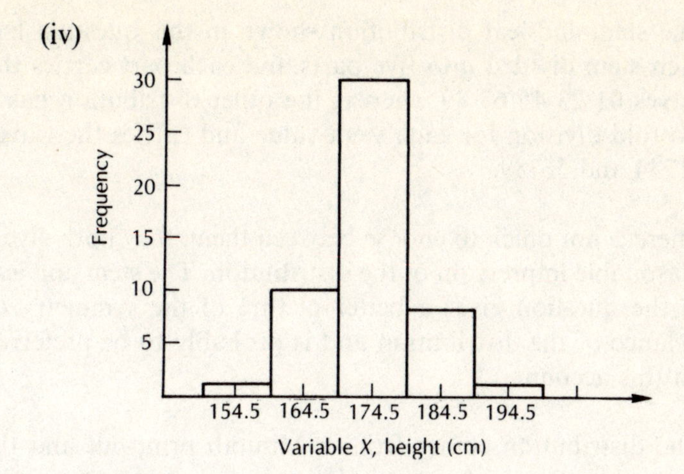

2 (i) 66-68 kgs

(ii)

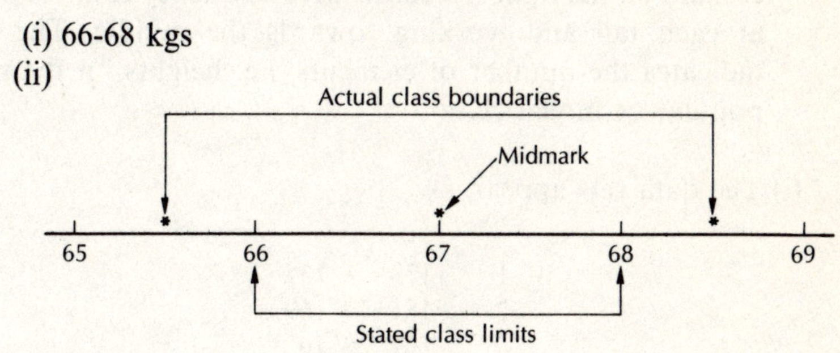

(iii) 3 kgs

(iv)

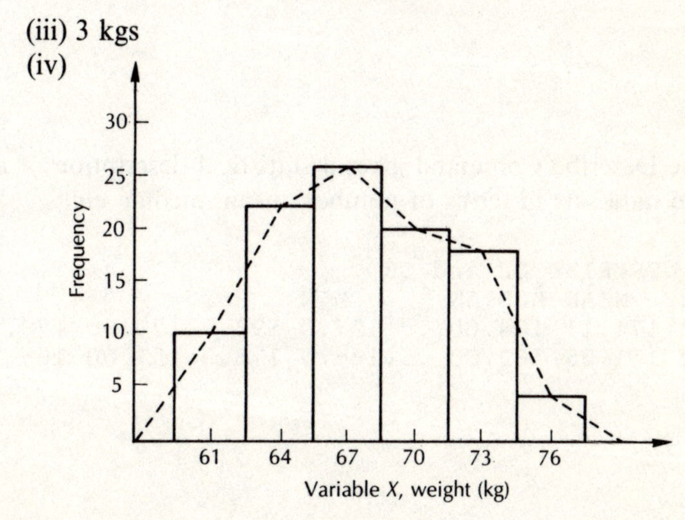

3 (i) The stem-and-leaf distribution shown in this question has each stem divided into five parts and each part carries the leaves 01 23 45 67 89 whereas the other distribution has a twofold division for each stem value and carries the leaves 01234 and 56789.

(ii) There is not much to choose between them; they both give a reasonable impression of the distribution. The stem and leaf in the question gives a better picture of the symmetry or balance of the distribution and is probably to be preferred on this account.

(iii) The distribution comes from a Minitab print-out and the column on the right is a cumulative frequency count starting at each tail and working towards the middle. The (14) indicates the number of elements, i.e. heights, in the most popular or modal class.

4 (i) The data sets appear so:

1	173	135
2	181	169
3	169	162
.	.	.
.	.	.
.	.	.

The Describe Command gives a statistical description of the two data sets in terms of number, mean, median etc.

```
MTB > DESCRIBE C2 and C3
      N   MEAN MEDIAN       MIN   MAX     Q1     Q3
C2 100 174.11 174.00 ...157.0 193.0 170.00 178.75
C3 100 161.25 162.00 ...143.0 176.0 157.00 165.75
```

(ii)

```
MTB > HISTOGRAM C2 C3
Histogram of C2 N = 100
Midpoint    Count
155          1 *
160          4 ****
165         10 **********
170         23 ***********************
175         35 ***********************************
180         18 ******************
185          5 *****
190          3 ***
195          1 *


Histogram of C3 N = 100
Midpoint    Count
144          2 **
148          2 **
152          7 *******
156         15 ***************
160         23 ***********************
164         26 **************************
168         15 ***************
172          8 ********
176          2 **
```

(iii)
```
MTB > STEM-AND-LEAF C2 C3
Stem-and-leaf of C2 N = 100
Leaf  unit = 1.0
   2    15   78
   7    16   02234
  23    16   5566667788899999
 (31)   17   0000011112222223333333344444444
  46    17   55555555666677777788899999
  20    18   00011111222334
   6    18   6688
   2    19   23
```

```
Stem-and-leaf of C3 N = 100
Leaf unit = 1.0
   1  14   3
   2  14   4
   2  14
   4  14  89
   8  15  0001
  11  15  333
  17  15  444555
  26  15  666667777
  36  15  8888899999
  49  16  0000000011111
(14)  16  22222223333333
  37  16  444444555555
  25  16  66667777
  17  16  8889999
  10  17  000001
   4  17  22
   2  17   4
   1  17   6
```

(iv) To get help while in Minitab you need to type HELP, Alt + H or use the F1 key. Or if you know what kind of help you want you can for instance type HELP OVERVIEW, HELP COMMANDS, HELP TABLE etc. In our case HELP HISTOGRAM and HELP STEM-AND-LEAF will suffice. As you can see the various methods of describing the distribution have a number of subcommands. Try out some of these commands if you have time and see if you can work out what they do. Don't spend too much time on this as we shall return to these subcommands later.

5 (i)
```
MTB > HISTOGRAM C7
Histogram of IMR N = 59
Midpoint  Count
     0      12  ************
    20      10  **********
    40       9  *********
    60       9  *********
    80       6  ******
   100       7  *******
   120       2  **
   140       2  **
   160       0
   180       2  **
```

(ii)
```
MTB > STEM-AND-LEAF C7
Stem-and-leaf of IMR N = 59
Leaf unit = 1.0
  12     0 567788888899
  17     1 00156
  22     2 02466
  25     3 019
  (6)    4 003689
  28     5
  28     6 011357889
  19     7 1556
  15     8 26
  13     9 49
  11    10 12244
   6    11 4
   5    12 3
   4    13 0
   3    14 4
   2    15
   2    16
   2    17 33
```

(iii)

```
MTB > DESCRIBE C7
           N     MEAN   MEDIAN   TRMEAN    STDEV  SEMEAN
IMR       59    55.03    48.00    51.68    43.88    5.71
         MIN      MAX       Q1       Q3
IMR     5.00   173.00    11.00    82.00
```

(iv) The distribution of the IMR data is very lopsided, i.e. it is skewed. For most of these 59 countries the IMR is found at the low end of the distribution. The minimum rate is 5, and this is to be found in Japan. From the stem-and-leaf diagram we can see that, of these 59 countries, 31 (i.e. 25 + 6 = 31) have an IMR of less than 50. That is, there are less than 50 annual deaths of infants under 1 year of age per 1000 live births in 1990. The median (middle) value in the distribution is 48 and occurs in Tunisia. There is a considerable tail extending to the highest values of 173 which are to be found in Angola and Mozambique. The distribution has a mean of 55 which is considerably greater than the median. This reflects the concentration of the data at the lower end of the scale. The spread of the data (as measured by the standard deviation) is also large.

4 Level and Spread

In this chapter we look at two types of measure or numerical summary with which we can describe a set of data or a distribution, measures of level (average) and measures of spread. However, before doing this we outline the sigma notation, which will help us describe the concepts used.

Measures of level indicate where the 'middle' of the distribution is. Where for instance does the average height lie in a set of data? Measures of spread tell us how the data are dispersed or spread around the 'middle' of the distribution. Are the heights, say, gathered tightly around this average to give a 'thin' or tightly grouped distribution or are they spread widely about the average to give a 'fatter' or more dispersed distribution? Such numerical measures can not only summarize, but more importantly can be used to compare different distributions.

We shall look at the following measures of *level* and *spread*;

Level	(i)	the median
	(ii)	the arithmetic mean
Spread	(i)	midspread
	(ii)	standard deviation

However, first we need to make ourselves familiar with a system of notation, the sigma notation, which we can use to clarify various characteristics of data sets and distributions.

The Sigma or Summation Notation

The sigma notation is a handy system that is much used in statistics. $\sum$ is the Greek capital letter sigma and is used here as an operator, i.e. it tells you to perform some operation on the data. We are all familiar with operators in the form of the symbols $+$, $-$, $\times$ and $\div$. They tell you to operate on the data in certain ways, add this to that, subtract this number from that, and multiply or divide numbers. $\sum$ is a similar operator but it indicates that you must add together or sum all the indicated items. For instance $\sum x$ simply means sum or add together all the x values. $\sum f$ means sum all the f values. $\sum fx$ means sum all the fx values.

To be more precise we need suffixes or subscripts to give some indication of where the summation will start and stop. Consider a simple data set, say the weights of nine judo players measured in kilograms to the nearest kilogram.

68 69 70 70 71 71 71 73 74

The variable here is weight and as usual we give it the symbol x. We have nine elements in the set and we will give each one a distinctive label or suffix.

68	69	70	70	71	71	71	73	74
x_1	x_2	x_3	x_4	x_5	x_6	x_7	x_8	x_9

We must be quite clear what these subscripts or suffixes are. They are merely labels, or addresses if you like. They indicate the value of the variable at a certain address or place. For instance x_2 (pronounced ecks-two) indicates the second value of the variable (69 kg) and so on.

In addition a more general suffix is used. x_i (pronounced ecks-eye) is a general symbol which stands for an unspecified x value. If we wish to sum the weights we may write

$$\sum x = x_1 + x_2 + x_3 + x_4 + x_5 + x_6 + x_7 + x_8 + x_9$$
$$= 68 + 69 + 70 + 70 + 71 + 71 + 71 + 73 + 74$$
$$= 637 \text{ kg}$$

To be more precise we should indicate the limits over which we wish to sum the data by the use of suffixes. For example. $\sum_1^9 x_i$ tells us to sum all the x_i values from x_1 to x_9 inclusive. This is equal to $x_1 + x_2 + ... + x_9 = 68 + 69 + ... + 74 = 637$ kg. Similarly, $\sum_1^4 x_i$ sums only the first four values from x_1 to x_4 inclusive and is equal to $x_1 + x_2 + x_4 = 68 + 69 + ... + 70 = 277$ kg. and $\sum_5^9 x_i = x_5 + x_6 + ... + x_9 = 360$ kg.

If we are dealing with frequencies the summation is done in a similar way but taking note that the data are now grouped together. Consider our simple data set again. We note that although there are nine elements in the set the variable x only takes on six different values and we can see that the summation can also be written as

$$\text{sum} = 1(68) + 1(69) + 2(70) + 3(71) + 1(73) + 1(74)$$
$$= 637 \text{ kg}$$

The figures in parentheses are the only six values that the variable (weight) takes and the figures outside the parentheses indicate the number of times or frequency with which each of these values occurs.

Consider the first term of the summation above, 1(68). This could be written in a more general form as $f_1(x_1)$ or f_1x_1 where f_1 indicates the frequency and x_1 the first value of the variable. We may now write the sum more generally as

$$\text{sum} = f_1x_1 + f_2x_2 + f_3x_3 + f_4x_4 + f_5x_5 + f_6x_6 = \sum fx$$

That is, the sum is equal to the value of the variable x multiplied by its corresponding frequency, i.e. x_1 is multiplied by f_1 to give f_1x_1, x_2 by f_2 to give f_2x_2 etc. (Note that the suffixes of f and x must always be the same; terms such as f_1x_3 are meaningless.)

As we have grouped the data we now have only six terms in our summation and we may write it as

$$\text{sum} = \sum_1^6 f_ix_i$$

We should also note that the order of the operation is crucial, first each x is multiplied by its corresponding frequency f and then these fx values are added up.

Although the above expression, adding the six fx terms, appears different from our first summation $\sum_1^9 x_i$, they both indicate the same quantity, 637 kg. The difference lies in the notation and occurs because we grouped the data in one and not the other. As we shall see later this will give us two apparently different formulae for both the mean and the standard deviation depending on whether we use the raw data or group it in some way.

After having stressed the suffix notation you may be surprised, but also relieved, to know that where there is no possibility of confusion regarding what is to be summed over what limits etc. we will leave the suffixes out and merely use the symbols $\sum x$ with raw data and $\sum fx$ with grouped data. To reinforce the ideas outlined above let us do a simple example. Although we do not strictly need suffixes here I have used them to illustrate part of the solution.

The following is a set of weights of 12 dogs measured to the nearest kilogram: 1, 4, 5, 2, 2, 0, 1, 2, 1, 3, 4, 1. Draw up a frequency distribution table and find the following:

$$\text{(i)} \sum f \qquad \text{(ii)} \sum fx \qquad \text{(iii)} \frac{\sum fx}{\sum f}$$

The variable x only has six values, i.e.

$$0 \quad 1 \quad 2 \quad 3 \quad 4 \quad 5 \text{ kg}$$

We can call these

$$x_0 \quad x_1 \quad x_2 \quad x_3 \quad x_4 \quad x_5$$

By counting we see that the corresponding frequencies are

$$1 \quad 4 \quad 3 \quad 1 \quad 2 \quad 1$$

and we call these

$$f_0 \quad f_1 \quad f_2 \quad f_3 \quad f_4 \quad f_5$$

So we get

$$\text{(i)} \sum f = f_0 + f_1 + \ldots + f_5 = 1 + 4 + \ldots + 1 = 12 = N$$

$$\text{(ii)} \sum fx = f_0x_0 + f_1x_1 + \ldots + f_5x_5 = 1(0) + 4(1) + \ldots + 1(5) = 26 \text{ kg}$$

(iii) $\dfrac{\sum fx}{\sum f} = \dfrac{26}{12}\text{kg} = 2.2 \text{ kg}$

$\sum f$ is the total frequency and is often given the symbol N and as we shall see later the expression $\sum fx/\sum f$ is defined as the arithmetic mean of a frequency distribution.

Another small point worthy of note is that one of the dogs has been given a weight of 0 kg. Is this possible? Yes indeed! We are measuring to the nearest kilogram so this little beast must have had a weight of less than half a kilogram. However, don't be tripped up; remember that multiplying by zero always gives zero. So $f_0x_0 = 1(0) = 0$ kg. So much for small dogs. Let us now look at measures of level (average).

Measures of Level

The median

The median is the value of the middle element of the variable in a data set or frequency distribution when all the elements are arranged in order of size. For example if we had 21 students in a seminar group and we wished to find the median height for the group, we could ask them to form one line with the shortest on the left and arrange themselves in order of height until the tallest person was standing on the right. The eleventh person from either end will have the median height of the group.

Or again, let us say we have a distribution of the ages of seven children in a nursery as follows, 1, 2, 2, 3, 4, 4 and 4 years. The median of this distribution is the middle value and can be seen to be 3 years. Half the distribution lies above and half below the median and it is for this reason that it is sometimes called the 50th percentile.

In both the above examples the number of elements being considered is an odd number, i.e. $N = 21$ and $N = 7$, and so in each case there is unambiguously a middle element. Using the suffix notation and starting at x_1 the medians will be the values of x_{11} and x_4 respectively. However, if there is no middle element, i.e. N is an even number, then the median is taken to be the arithmetic mean of the two middle elements. For example if an eighth child joined the nursery and her age was 4 years, the distribution would now be

1	2	2	3	4	4	4	4 years
x_1	x_2	x_3	x_4	x_5	x_6	x_7	x_8

The median is taken to be halfway between x_4 and x_5, i.e.

$$\tfrac{1}{2}(x_4 + x_5) = \tfrac{1}{2}(3 + 4) = 3.5 \text{ years}$$

Exactly half the distribution lies on either side of this point.

In more general terms the median and other dividers of the distribution could be illustrated as follows. Consider a simple symmetrical distribution of a variable as illustrated in figure 4.1. By symmetrical I mean that the variable is spread evenly about its middle point.

x_m is the median or the 50th percentile and so has 50% of the distribution either side of it. The area shaded represents the 50% of the distribution which is less than x_m.

By extending this thinking we can get two other percentile values. x_{ql} is the 25th percentile or lower quartile and divides the distribution or area under the 'curve' and the x axis into two pieces representing 25% of values less than it and 75% more than it. Similarly x_{qu} is the 75th percentile or upper quartile and has 75%

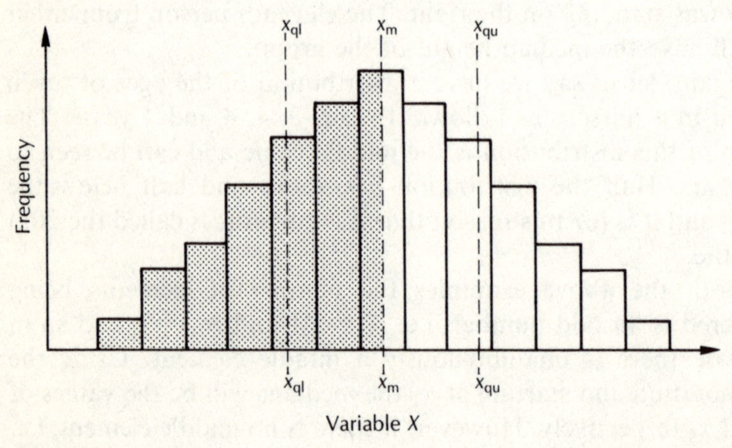

Figure 4.1

and 25% of the area under the curve either side of it. We will use these quartile values later when we look at measures of spread but let us now return to finding the median.

Finding the median is easy once the data have been ordered or grouped according to size. Consequently it is very easy to find the median, or indeed other quartiles, from a stem-and-leaf distribution. All we need to do is to count in halfway from either end, or tail, of the distribution and we find the median. One small problem arises as we have seen from a consideration of the simple distributions considered earlier: how far do we count in? As we know, we have to find the middle point, the 50th percentile, which splits the distribution exactly in two halves. Half the distribution lies above this value and half below it.

This is done most conveniently by an approximation. Consider N, the total number of elements in the distribution. The median is found by counting in $\frac{1}{2}(N + 1)$ elements from either end of the distribution as we did for for the ages of the seven nursery children ($N = 7$). The median was found at $\frac{1}{2}(N + 1) = \frac{1}{2}(7 + 1) = 4$, i.e. at the fourth value from either end (3 years).

Let us look at another couple of examples.

Example 1 The IQ scores for 30 children were found to be as follows:

100	122	103	112	95	97
115	107	101	104	98	115
107	115	88	110	87	92
121	100	101	105	112	101
107	105	128	108	95	94

Construct a stem-and-leaf distribution of the data and find the median IQ score.

Using stems of tens we get

	Stem	Leaf
2	8	78
8	9	245578
(13)	10	0011134*557778
9	11	022555
3	12	128

Stem: tens Leaf: units $N = 30$

You will notice that alongside the distribution I have given a column which is a cumulative count in from each end in the manner that Minitab adopts. The figure in parentheses indicates the most popular or modal class and our median x_m lies in here.

As $N = 30$ we want the $\frac{1}{2}(N + 1)$ value, i.e. $\frac{1}{2}(30 + 1) = 15.5$. Counting into the 15.5 place we get an IQ score of 104.5 (indicated by an asterisk).

Example 2 Let us now look at the set of women's heights we used in chapter 3. We will find the median x_m and to give us some counting practice we will also find the lower (x_{ql}) and upper (x_{qu}) quartiles.

	Stem	Leaf
2	14	34
4	14	89
14	15	0001333444
36	15	55566666777*78888899999
(33)	16	000000001111 12*222222333333444444
31	16	555555*666677778889999
10	17	000001224
1	17	6

Stem: tens Leaf: units $N = 100$

(the asterisks mark the approximate median and quartile values.)

On the left-hand side of the distribution I have again given a cumulative count from each end of the distribution. (33) indicates the modal class. As $N = 100$,

x_m is found at the halfway mark,

$$\tfrac{1}{2}(N + 1) = \frac{1 \times 101}{2} = 50.5\text{th place}$$

x_{ql} is found at the quarterway mark,

$$\tfrac{1}{4}(N + 1) = \frac{1 \times 101}{4} = 25.25\text{th place}$$

x_{qu} is found at the threequarter mark,

$$\tfrac{3}{4}(N + 1) = \frac{3 \times 101}{4} = 75.75\text{th place}$$

Note, however, that the upper quartile x_{qu} is more easily found by counting in 25.25 places from the higher end of the distribution. So counting from the ends of the distribution we get

$$x_m = 162 \text{ cm (median)}$$
$$x_{ql} = 157 \text{ cm (Q1)}$$
$$x_{qu} = 165.75 \text{ cm (Q3)}$$

Although the focus of our interest in this section is with the median, it is worth noting that the upper and lower quartiles can be found from stem-and-leaf distributions by the same approximate counting technique. For the median we go in to the halfway mark but for the quartiles we only go a quarter of the way into the distribution from each end. In the above case it should be noted that a bit of interpolation is necessary to find x_{qu} accurately.

It is also worth noting that the lower (x_{ql}) and upper (x_{qu}) quartiles in Minitab are given the labels Q1 and Q3 respectively. One might expect that logically the median x_m would be labelled Q2, the second quartile; however, this is not the case in Minitab although some writers do use the term.

The arithmetic mean

The arithmetic mean is one of the most important measures of average, location or level. It is just the common or garden average and, as we have seen earlier, can be found by adding all the values of the variable and dividing by the number of variables in the set or distribution.

Other types of mean also exist, such as the geometric mean and the harmonic mean, but we are not going to consider these and as there is no risk of confusion we will refer to the arithmetic mean simply as the mean and give it the symbol $\bar{X}$ (pronounced bar-ecks). There are many ways of calculating it but here we shall just look at two straightforward methods, one using the raw data and the other grouped data. We will use the notation developed previously and give a formula for each case.

Consider the weights of 10 girls in kilograms measured to the nearest kilogram: 59, 60, 61, 61, 61, 61, 62, 62, 62, 69 kg. It is quite easy to find the mean $\bar{X}$ by adding up the weights and dividing by 10, the number of girls ($N = 10$).

$$\text{mean} = \bar{X} = \frac{59 + 60 + \ldots + 69}{10} = \frac{61.8}{10} = 61.8 \text{ kgs}$$

Note that the mean has the same dimension as the variable, in this case, weight in kilograms.

Let us now look at the mean more generally, first using the raw data and then grouping it.

The variable can be labelled as follows:

59	60	61	61	61	61	62	62	62	69	(N = 10)
x_1	x_2	x_3	x_4	x_5	x_6	x_7	x_8	x_9	x_{10}	

The mean can then be defined more formally as

$$\bar{X} = \frac{\sum_1^{10} x_i}{N} = \frac{x_1 + x_2 + \ldots + x_{10}}{10} = \frac{59 + 60 + \ldots + 69}{10} = 61.8 \text{ kg}$$

If we group the data by taking frequency into account we only need to consider five values of the variable weight, i.e. 59, 60, 61, 62 and 69 kg, and by multiplying the variable by the appropriate frequency we get

$$\bar{X} = \frac{1(59) + 1(60) + 4(61) + 3(62) + 1(69)}{1 + 1 + 4 + 3 + 1} = 61.8 \text{ kg}$$

more generally this can be expressed as

$$\bar{X} = \frac{f_1x_1 + f_2x_2 + f_3x_3 + f_4x_4 + f_5x_5}{f_1 + f_2 + f_3 + f_4 + f_5} = \frac{\sum_1^5 f_ix_i}{N}$$

The two formulae (1) and (2) can be seen to be equivalent and where there is no danger of confusion regarding limits of summation etc. we can express them in their simple form.

$$\text{mean} = \bar{X} = \frac{\sum x}{N} \quad \text{for raw data}$$

$$\text{mean} = \bar{X} = \frac{\sum fx}{N} \quad \text{for frequency distributions}$$

The mean can be quite easily found by using one of these formulae as appropriate. Let us consider a simple frequency distribution of heights. We will find the mean using a tabular layout and a pocket calculator and then I shall indicate how it can be done just using the statistical functions of the calculator.

Example 3 The following is a frequency distribution of the heights of 50 randomly selected first-year female undergraduates measured to the nearest inch.

Height in inches	*Frequency*
60–62	4
63–65	15
66–68	21
69–71	9
72–74	1

Calculate the mean height of the sample.

1 As it is a frequency distribution we will of course use the formula $\bar{X} = \sum fx/N$; however, as the data are given here we only have groups or classes but no x values. We will make the usual assumption that the data in each class are grouped symmetrically about the middle of each class and so use the class midmarks as x values. Laid out in tabular form we then get the following.

Class	*Midmark (x)*	*Frequency (f)*	*fx*
60–62	61	4	244
63–65	64	15	960
66–68	67	21	1407
69–71	70	9	630
72–74	73	1	73
	Total	50	3314

The entries in the fx column are obtained by multiplying the x values by their corresponding frequency, using a pocket calculator as necessary, i.e.

$$f_1x_1 = 4 \times 61 = 244, \ f_2x_2 = 15 \times 64 = 960 \ etc.$$

$$\text{mean} = \bar{X} = \frac{\sum fx}{N} = \frac{3314}{50} = 66.3 \text{ inches}$$

2 Using a pocket calculator with statistical functions: it is first necessary to put the calculator into statistical mode and then

punch in the information. How exactly this is done will of course vary according to the make of the calculator. However, I shall give a brief outline of the procedure using the popular Casio *fx* models. Many other makes follow similar patterns but you should consult your calculator manual for details.

(a) Clear the memories by ϕ Min , switching the power off and on, or otherwise.
(b) Press inv mode until SD appears on the screen to indicate that you are in statistical mode.
(c) Looking at the distribution as laid out in (1) enter the data in this order and store it in the memory with M+, the add to memory key.

    61 × 4 M+ (i.e. $x_1 \times f_1$ M+)

    64 × 15 M+

    etc. until

    73 × 1 M+

    Note: The order does matter: x precedes f and there is no = sign in these keying sequences.
(d) Pressing N should now give 50, and a mean of 66.28 inches (66.3 inches to one decimal place)

You could ask yourself what the other statistical function buttons indicate and it is certainly worth noting that σ_n gives a value of 2.72. This is a measure of spread called the standard deviation which we shall look at shortly. However, first we need to say a few words about the uses and characteristics of the two measures of average we have examined.

Using the Median and Mean

Now we know how to find the value of these measures of average we need to give some thought to when it is appropriate to use them and when not.

The median and the mean are descriptive statistics and as the name implies they describe the average or level of a data set or distribution. Also as single number summaries of level they are useful for the comparison of distributions.

The mean is the more commonly used measure. It is easy to find, uses all the data and its frequent use, and indeed misuse, in the everyday economic and social world gives us a more intuitive feel for its meaning. Within statistics it is also the most common measure of level. It is closely linked to standard deviation which measures spread and these two descriptive statistics can together be used to describe many populations with known precision. However, we must be aware that the mean (and indeed the median) is not always an appropriate measure. Before calculating and using either we need to ask if it is meaningful.

The median, in that the data have to be ordered into a stem-and-leaf plot, takes longer to find, unless of course we have a computer and program to hand with the data nicely entered into a suitable database. However, with small sets, such as we have looked at, the ordering and counting does not take long and we get the considerable bonus of a rough sketch of the shape of the distribution. The weakness of the median, however, is that it wastes so much information. Leaving aside the value of the stem-and-leaf distribution for the moment, just looking at the median we can see that the only values that matter at all are the middle or two middle values. Once the data are ordered and stay ordered, changes in any or all of the other values will have no effect on the value of the median. This reliance on one or two middle values, in the absence of a stem-and-leaf or other plot of the distribution, can in some cases be a great disadvantage and mislead us.

However, in other cases we should note that such disregard of all other values may be an advantage. Its weakness becomes its strength. If we are dealing with data that are ragged or incomplete or data that are in some way suspect or not as clean as we would wish, then such indifference to all values except the middle ones may be an advantage. A great many interesting databases do have parts where the data are incomplete or of suspect accuracy and validity. For such data the median might well be the appropriate measure of level to use at least for initial explanation. It is what is called a resistant measure of level. Errors, wild or incomplete observations, extremes or freaks at the ends of the distribution will affect the median less than the mean.

The median as a resistant measure of level is much used in pilot studies and exploratory data analysis. The mean, by contrast, with its stable-mate the standard deviation, is greatly affected by odd or unusual values and may consequently provide us with a misleading summary of the distribution. It uses all the data and gives equal weight or importance to each element; consequently unusually large values, for instance, can have considerable and perhaps damaging effect.

We could illustrate one way in which the mean is capable of misleading by considering, say, the ages of 10 students in a university seminar group. Let us say that the ages were 18, 18, 18, 18, 20, 20, 20, 21, 21 and 46 years. We can calculate the mean age of these students to be 220/10 = 22 years. This is obviously not a good or representative measure of the ages in the group. If we did not have the data before us we might accept this as a reasonable description of the group ages: however, when we look at the data we see how poor the description provided by the mean is. Nine out of ten people in the group are younger than the mean and the remaining person, at 46 years of age, is more than twice as old. In this case it is clear that the measure of level chosen is inappropriate and misleading.

The median of the data can be seen to be 20 years and this gives us a somewhat better, but still very partial, picture of the data set. This measure, of course, takes no account of the extreme value of 46 years which made the mean so unrepresentative. We should also note that the mature student could in fact be much older than 46 and exert not the slightest effect on the median.

The message here is clear. We need to exercise care in our choice of a statistic to represent level. The median as the more resistant measure is to be preferred for exploratory work, especially with uncertain data. The mean uses all the data but may be misleading for lopsided or skewed distributions or where there are outliers or unusual values. The mean is used almost exclusively, however, in confirmatory analysis and especially in statistical inference where a great deal of theory has been developed to enable us to infer characteristics of a whole population by considering only part of it.

Finally, we should also note that one type of statistic describing level (average), whether it be the mean or median, cannot by itself give us an adequate picture. We also need, at minimum, a measure of spread. We now look at two measures of spread.

Measures of Spread

After measures of level or average the most common characteristic used to describe a distribution is spread. Many distributions can be described adequately in terms of level and spread.

Looking at the stem-and-leaf plot of a distribution which is unimodal, i.e. which has one hump such as the distributions we looked at in chapter 3, what strikes the eye first is where the centre or middle of the distribution lies. The measures of level we have just looked at can quantify this. What then strikes one is how the data are distributed or spread about this level or centre. The eye takes in how 'fat' or 'thin' the distribution is and whether there are, say, long tails at the ends of the distribution. What we are talking about here is dispersion about the average value and we shall look at just two statistics which measure it, the midspread and the standard deviation.

These two measures of spread are paired with the measures of level (average) that we looked at earlier. As you can guess the midspread, or quartile range as it is sometimes called, goes with the median. The median and midspread are both resistant measures and are used primarily in exploratory data analysis. The standard deviation is paired with the mean and although neither is resistant, being easily corrupted by outliers or freak values, they are two of the most common and important measures in confirmatory statistics. They form much of the core of the statistics of inference.

Earlier, in chapter 3, we did come across a crude measure of spread, the range. As you may remember the range of a variable is the difference between the largest and smallest values of the variable. For example in the data set of women's heights (table 3.1) the tallest woman is 176 cm (say x_u) and the shortest 143 cm (say x_l) and so the range is $x_u - x_l = 176 - 143 = 33$ cm. This measure of spread is both easy to understand and simple to calculate; however, it is not much used as it suffers from some obvious defects. It is not a resistant measure as it is dependent only on the two extreme values in the distribution and these are often the very ones that are most erratic. Only one 'giant' or 'dwarf' in a large population would be needed to give a totally inaccurate picture of the spread in the distribution.

We need to go in some way from the possibly erratic ends of the distribution to form our first measure of spread. The midspread goes in a quarter of the way from each end of the distribution.

The midspread (inter-quartile range)

The midspread is similar to the range but we measure it not between extreme values but between the quartile values which are embedded in the more stable central part of the distribution. The midspread is the difference between the upper (x_{qu}) and lower (x_{ql}) quartiles. It is usually given the symbol d_q, standing for the difference between the quartiles.

$$\text{midspread} = d_q = x_{qu} - x_{ql}$$

If you refer back to the distribution of women's heights (chapter 3) you will see that by counting we found the median and the two quartiles to be

$$x_m = 162 \text{ cm}$$
$$x_{qu} = 165.75 \text{ cm}$$
$$x_{ql} = 157.00 \text{ cm}$$

Therefore midspread $d_q = 165.75 - 157.00 = 8.75$ cm. The midspread is easy to calculate once we have ordered the data into a stem-and-leaf distribution and counted in to find the quartile values. We can summarize the various values as follows:

N number of elements in the distribution of the variable x
x_u upper or maximum value of the variable x
x_{qu} upper quartile value of the variable x (Q3)
x_m median value of the variable x (Q2)
x_{ql} lower quartile value of the variable x (Q1)
x_l lower or minimum value of the variable x
d_q midspread of the variable x

This information is conveniently illustrated by a boxplot (box-and-whisker diagram). Minitab can produce a variety of boxplots using various subcommands. However, let us look first at a simple boxplot drawn by hand using the women's height data.

Looking back at the data and the relevant stem-and-leaf distribution we can summarize as follows:

N	100
x_u	176 cm
x_{qu}	165.75 cm
x_m	162 cm
x_{ql}	157 cm
x_l	143 cm
d_q	8.75 cm

From this summary a boxplot can be constructed as in the upper part of figure 4.2. We note that the box is drawn between the two quartiles and contains the median and the middle half of the distribution. The whiskers extend each side to the extreme values. However, in the lower part of the figure I have shown a boxplot of the same data as it could appear in a Minitab output.

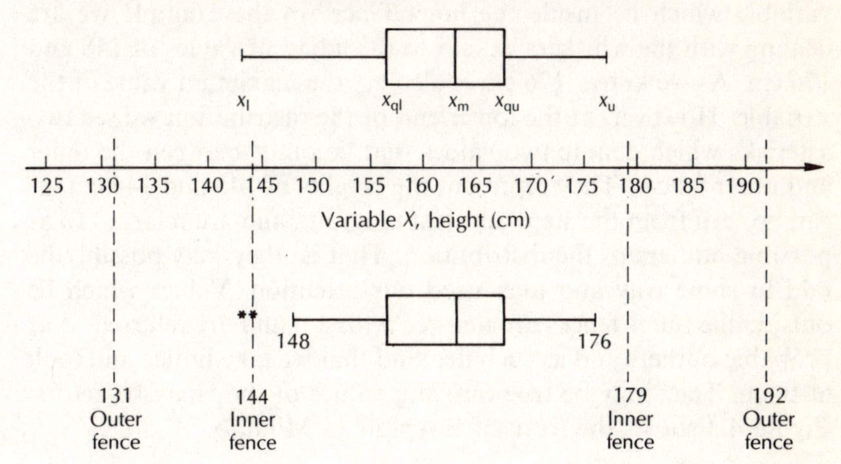

Figure 4.2

We need to note a few points regarding this Minitab boxplot. The box or centre half of the distribution is in the same place but we note that now only one whisker extends to x_u, the height of the tallest woman. The other whisker stops well short of the height of the smallest woman and two asterisks appear after the end of the whisker. To understand this we need to look briefly at what are described as fences and how these are found. If you require a full description you should refer to the Minitab manual.

Inner and outer fences are calculated on each side of the distribution. Minitab, uses the concept of hinge, which for all practical exploratory purposes can be taken to be equivalent to the quartiles, so the position of the inner fence is found as follows (to the nearest centimetre):

$$x_{ql} - 1.5d_q = 157 - 1.5(8.75) = 144 \text{ cm}$$
$$x_{qu} + 1.5d_q = 165.75 + 1.5(8.75) = 179 \text{ cm}$$

The position of the outer fence is found in a similar manner but using three times the midspread in the formulae in place of 1.5, to give 131 and 192 cm.

The details of these calculations are unimportant but data values outside the fences i.e. outliers, always need to be noted.

The whiskers on either side of the box are drawn to what are called the adjacent values; these are the most extreme values of the variable which lie inside the inner fence. In the example we are dealing with the whiskers extend to the adjacent values of 148 and 176 cm. As we know, 176 cm is also x_u, the maximum value of the variable. However, at the lower end of the distribution we see two asterisks which denote two values that lie on or between the inner and outer fences. They represent the heights of 143 and 144 cm, as can be seen from the stem-and-leaf diagram, and are referred to as possible outliers of the distribution. That is, they may possibly be odd in some way and may need our attention. Values which lie outside the outer fences are marked with a o and are referred to as probable outliers and as such demand that we take notice and look at them. They may be true outlying values or they may be errors. Figure 4.3 shows the form of Boxplots in Minitab.

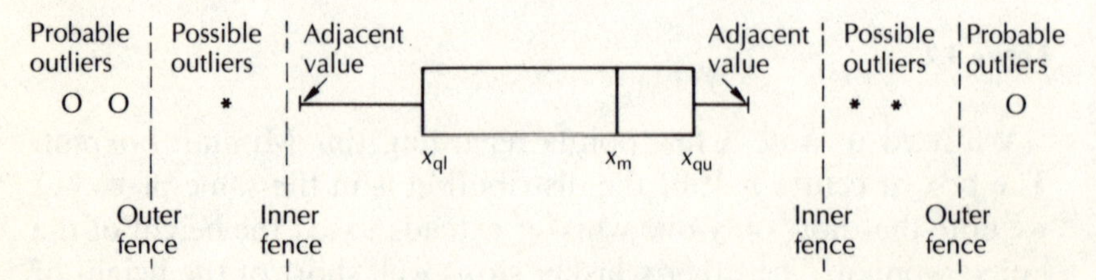

Figure 4.3

We shall return to boxplots in later chapters but now we look at another measure of spread, the standard deviation.

The Standard Deviation

The standard deviation measures the spread of the data around the arithmetic mean $\bar{X}$. For raw or ungrouped data it is defined as follows:

$$\text{standard deviation} = s = \sqrt{\left[\frac{\sum(x - \bar{X})^2}{N}\right]} \qquad (1)$$

For grouped data we have to take frequency into account to get

$$\text{standard deviation} = s = \sqrt{\left[\frac{\sum f(x - \bar{X})^2}{N}\right]} \qquad (2)$$

where $\bar{X}$ is the arithmetic mean, x the variable, f the frequency and N the total number of data. For many purposes, especially when dealing with samples, we shall make the minor modification of using $N - 1$ instead of N as denominator. It is easier, however, to get an intuitive feel for the concept of standard deviation by looking first at the definitional formula for raw data given above and looking at some simple examples. After this we shall look at some grouped data and calculate the standard deviation by hand and then by using the statistical functions of a pocket calculator.

First let us look at formula (1). At the heart of this formula is the quantity $(x - \bar{X})$. This tells us that we need to subtract the mean $\bar{X}$ from every value of the variable x that we have in the data set. As we can see from the sigma sign $\sum$, we need to add up or sum all these differences or deviations from the mean. However, as the mean $\bar{X}$ is a middle value in the distribution it will be bigger than some values of x but smaller than others. That is, sometimes $(x - \bar{X})$ will be negative and sometimes positive. As we wish to sum the differences from the mean we need to ensure that the negative and positive differences from the mean do not cancel each other out. We do this by squaring all the differences $(x - \bar{X})$ to give a positive measure of the deviation from the mean. Remember the convention in mathematics: if we multiply a negative quantity by itself we get a positive quantity, e.g. $(-3) \times (-3) = +9$.

We now sum all these squared differences from the mean to get $\sum(x - \bar{X})^2$. If we divide this by N, the number of elements in the data set, we get a measure of the 'average' squared deviation from the mean. We may now, not quite correctly, see the square root sign as 'cancelling out' the squaring of the differences undertaken earlier.

Let us illustrate the concept of standard deviation by two examples using simple numbers. Consider first three small distributions of the ages of children in three families:

(a) 5.0 5.0 5.0 years
(b) 3.0 5.0 7.0 years
(c) 2.0 5.0 8.0 years

Describe the distribution of childrens' ages in each family by finding the mean and standard deviation.

The mean, in each case, is 5 years. This can be found by inspection, that is, by looking at the 'balance' of the variables or by adding all the values together and dividing by 3. For example, the mean of distribution (b) is $\bar{X} = (3+5+7)/3 = 5.0$ years.

Let us now look at the standard deviation of each distribution. Using the definitional formula

$$\text{standard deviation} = s = \sqrt{\left[\frac{\sum(x-\bar{X})^2}{N}\right]} \quad (1)$$

we get the following three standard deviations,

$$s = \sqrt{\frac{(5-5)^2 + (5-5)^2 + (5-5)^2}{3}} = 0 \text{ years}$$

$$s = \sqrt{\left[\frac{(3-5)^2 + (5-5)^2 + (7-5)^2}{3}\right]} = \sqrt{\left(\frac{4+0+4}{3}\right)}$$

$$= \sqrt{2.66} = 1.6 \text{ years}$$

$$s = \sqrt{\left[\frac{(2-5)^2 + (5-5)^2 + (8-5)^2}{3}\right]} = \sqrt{\left(\frac{9+0+9}{3}\right)}$$

$$= \sqrt{6} = 2.4 \text{ years}$$

The three simple distributions have the same mean, 5 years, but standard deviations of 0, 1.6 and 2.4 years. The standard deviation measures the spread of the data about the mean of 5 years, and in the first case we see that it is zero, i.e. there is no spread as all the observations have the same value of 5 years. Generally, however, the standard deviation will be greater than zero. The wider the dispersion or spread of data about the mean, the greater will be the standard deviation.

Let us now look at another simple example, this time taking frequency into account.

Example 4 Find the mean and standard deviation of the following distribution of the ages of 10 children in a nursery: 2, 2, 3, 4, 4, 4, 4, 5, 6, 6 years. Using the formulae for grouped data we get

$$\text{mean } \bar{X} = \frac{\sum fx}{N} = \frac{2(2) + 1(3) + 4(4) + 1(5) + 2(6)}{2+1+4+1+2}$$

$$= \frac{4+3+16+5+12}{10} = 4 \text{ years}$$

$$\text{standard deviation} = s = \sqrt{\left[\frac{\sum f(x-\bar{X})^2}{N}\right]}$$

$$= \sqrt{\frac{2(2-4)^2 + 1(3-4)^2 + \ldots + 2(6-4)^2}{2+1+4+1+2}}$$

$$= \sqrt{(18/10)} = 1.3 \textit{ years}$$

The above computation is best done using a tabular layout as follows,

x	f	fx	$x-\bar{X}$	$(x-\bar{X})^2$	$f(x-\bar{X})^2$
2	2	4	−2	4	8
3	1	3	−1	1	1
4	4	16	0	0	0
5	1	5	1	1	1
6	2	12	2	4	8
totals	10	40			18

$$\bar{X} = \frac{\sum fx}{N} = \frac{40}{10} = 4 \text{ years}$$

$$s = \sqrt{\left[\frac{\sum f(x-\bar{X})^2}{N}\right]} = \sqrt{\frac{18}{10}} = 1.3 \text{ years}$$

You should compare the tabular calculation with the same calculation laid out 'algebraically'. Except for simple cases, or of course for illustration as above, the tabular layout is to be preferred.

The basic definitional formulae for ungrouped and grouped data can give us an insight into the concept of standard deviation but they are generally inconvenient for computation except in simple distributions or where, as above, the mean is a whole number. In many cases a computational formula derived from the basic frequency formula is used. I have derived one such formula for you at the end of this chapter. It is as follows:

$$\text{standard deviation squared} = \text{variance} = s^2 = \frac{\sum fx^2}{N} - (\bar{X})^2$$

The variance is merely the square of the standard deviation. A simple example will make the use of this derived formula clear.

Example 5 The following is a set of the weights of 12 men measured to the nearest kilogram:

69 71 69 70
70 69 71 72
72 71 73 69

Find the mean and standard deviation.

x	f	fx	fx^2
69	4	276	19044
70	2	140	9800
71	3	213	15123
72	2	144	10368
73	1	73	5329
Totals	12	846	59664

$$\bar{X} = \frac{\sum fx}{N} = \frac{846}{12} = 70.5\ kg$$

$$s^2 = \frac{\sum fx^2}{N} - (\bar{X})^2 = \frac{59664}{12} - (70.5)^2$$

$$= 4972 - 4970.25 = 1.75$$

which is the variance, and so the standard deviation $s = \sqrt{1.75} = 1.3$ kg.

As can be seen this computational formula also involves a considerable amount of work, especially if you do not have a calculator to hand. Let us now look briefly at the calculation of the standard deviation using the statistical functions of a pocket calculator. We use the same distribution of weights as above.

The procedure is as I described it earlier in this chapter. First we must clear the memories and put the calculator into statistical mode. The data can now be entered in either its raw form

69 M+ 71 M+ 69 M+ 70 M+ etc. until 69 M+

or as grouped data:

69 × 4 M+
70 × 2 M+

etc. until

73 × 1 M+

The order of entry does not matter for the raw data but, for the data grouped by frequency, order is crucial. That is, x precedes f and there is no = sign in the keying sequences. M+ indicates the Memory Plus Key.

After having keyed in the data in one form or another, press N to see that the correct number of data elements have been entered. We get

$$N = 12 \qquad \bar{X} = 70.5 \qquad \sigma_n = 1.3228 \qquad \sigma_{n\text{-}1} = 1.3817$$

Note that the standard deviation which has been given the symbol lower case sigma, σ, is given in two forms, one using N and one using $N-1$ as denominator.

This is the time to remind you of a point I made earlier. For many purposes especially when we are dealing with sample data we will use the formula (for ungrouped data)

$$\text{standard deviation} = s = \sqrt{\left[\frac{\sum (x - \bar{X})^2}{N - 1}\right]}$$

It is this sample standard deviation formula that Minitab uses to calculate the standard deviation.

Derivation of a Computation Formula from the Basic Definitional Formula (If you are not mathematically inclined or your algebra is too rusty for you to appreciate this, don't worry – skip it and carry on with the exercises.)

By definition

$$s^2 = \frac{\sum f(x - \bar{X})^2}{N} = \frac{\sum f(x - \bar{X})\ (x - \bar{X})}{N}$$

Multiplying

$$s^2 = \frac{\sum f(x^2 - 2x\bar{X} + \bar{X}^2)}{N}$$

$$= \frac{\sum fx^2}{N} - \frac{\sum f(2x\bar{X})}{N} + \frac{\sum f\bar{X}^2}{N}$$

$$= \frac{\sum fx^2}{N} - \frac{2\bar{X}\sum fx}{N} + \frac{\bar{X}^2\sum f}{N}$$

2, $\bar{X}$ and $\bar{X}^2$ are constants like a, and $\sum ax = a\sum x$.

$$\frac{\sum fx}{N} = \bar{X}$$

and

$$\sum f = N$$

and so

$$\frac{\sum f}{N} = \frac{N}{N} = 1$$

$$s^2 = \frac{\sum fx^2}{N} - 2\overline{XX} + \overline{X}^2$$

$$= \frac{\sum fx^2}{N} - \bar{X}^2$$

Postscript

In this chapter we have examined two kinds of measure which can help us summarize or describe a distribution of data, measures of average or centrality and measures of spread or dispersion. The mean and median are measures of average and tell us where a distribution of data is positioned or centred. The way the data are grouped about these central or average values can be summarized by the second type of measure, spread. These paired measures of average and spread are fundamental to data analysis and will form an important ingredient of the work in subsequent chapters.

Exercises

1 The following is an array of the number of children in 12 families: 2, 4, 1, 5, 2, 0, 2, 2, 4, 2, 1, 4. Construct a frequency distribution table where x is the variable and f the frequency and find the following:

(i) $\sum f$ (ii) $\sum fx$ (iii) $\sum fx/N$

2 The basic weekly pay in pounds sterling of four men in an oil-rig gang were 100, 120, 140 and 600. What are the mode (the value of the variable that occurs most often), median and mean weekly pay of the group? How representative is the mean? (It is of course necessary to write a few sentences here.)

3 What are the median, mode (the value that occurs most often) and mean of this set of 10 numbers? 2, 2, 3, 5, 5, 5, 6, 6, 8, 9
If the median, mode and mean of distribution are equal or nearly equal what can we say about the distribution?

4 Calculate the mean and standard deviation of the following two distributions using the definitional formulae (denominator N):

(i) 5, 7, 9
(ii) 3, 7, 11

Check your answer using your pocket calculator in statistical mode.

5 Subtract 2 from each value in the distributions given in exercise 4 to give

(i) 3, 5, 7
(ii) 1, 5, 9

Find the mean and standard deviation of these two new distributions using your pocket calculator in statistical mode. Write a few sentences to explain the differences and similarities in the statistics for the old and new distributions (denominator N).

6 By selecting from the numbers 0, 1, 2, 3, 4, 5, 6, 7, 8, 9 write down
(i) four numbers which have the largest possible mean. What is their standard deviation?

(ii) four numbers which have the smallest possible mean. What is their standard deviation?
(iii) four numbers which have the largest possible standard deviation. What is their mean and standard deviation?
(iv) four numbers which have the smallest possible standard deviation.

In making your selection each integer (whole number) in the original set of numbers may be used more than once. i.e. this is sampling with replacement.

7 Using a tabular layout calculate the mean, variance and standard deviation (denominator N) of the following distribution:

x	1	2	3	4	5
f	2	1	4	1	2

8 The numbers of children of 10 women in a 40–44 year cohort were as follows:

number of children	0	1	2	3	4	5
frequency	0	1	2	5	0	2

Use your calculator in statistical mode to calculate the mean and standard deviation of the distribution.

9 Illustrate the distribution of 100 men's heights in table 3.1 using a stem-and-leaf plot. Use a stem of 10 units and split it in two parts, i.e. leaves of 0, 1, 2, 3, 4 and 5, 6, 7, 8, 9.

By counting find (i) the median and (ii) the upper and lower quartiles.

10 In table 3.2 you will find a frequency distribution of the heights of 100 men. From it calculate the arithmetic mean and standard deviation of the distribution using a tabular layout. (Use the standard deviation formula with N as denominator. You will need to use your pocket calculator here in non-statistical mode.)

11 Using your calculator in statistical mode find the mean and the standard deviation of the heights of 100 men from table 3.2 using both N and $N-1$ as denominator.

12 In this exercise we use the data file DATA02.DAT. It contains the heights of 100 men and 100 women. Call up the Minitab package and read in the data file using the command MTB > READ 'A:\DATA02.DAT' into C1-C3. Name columns C1 and C2 in the Editor as HTMEN and HTWOMEN respectively. Use the OUTFILE and NOOUTFILE commands at the beginning and end of the analysis in order to save your analysis. For example you could OUTFILE your analysis with the command MTB > OUTFILE 'A:\SARAH02'.

(i) Use the DESCRIBE command to get a statistical description of the distributions.

(ii) Boxplot the two distributions of heights by using the commands
MTB > BOXPLOT C2
MTB > BOXPLOT C3

(iii) Stem and leaf the two sets with the commands
MTB > STEM-AND-LEAF C2
MTB > STEM-AND-LEAF C3
Compare the boxplots and stem-and-leaf distributions and identify the values in each set which are indicated as possible outliers. Compare the median and quartiles given in the statistical description (i) with the medians and ends of the boxes in the plots.

(iv) Use the Minitab HELP facility to get information regarding boxplots using the command MTB > HELP BOXPLOT. Read it through and note what it says about quartiles, hinges, fences, outliers and adjacent values.

Answers to Exercises

1

x	f	fx
0	1	0
1	2	2
2	5	10
3	0	0
4	3	12
5	1	5
Totals	12	29

(i) $\sum f = N = 12$
(ii) $\sum fx = 29$
(iii) $\sum fx/N = 29/12 = 2.4$ children

This is the arithmetic mean number of children per family.

2 Median £130; there is no mode. The mean pay is £240, but this has little meaning here. The mean is obviously a poor representative statistic. Three of the men are well below this mean and one man earns more than twice as much as the mean. This example shows that we need to choose our representative statistic with care. The number in the 'sample' is ludicrously small, however.

3 Median 5 units; mode 5 units; mean 5.1 units. If median, mode and mean are equal or nearly equal the distribution shows symmetry. That is, it is fairly well balanced about the centre.

4 (i) $\bar{X} = 7$ units $N = 3$ $\sigma_n = 1.633$ units
(ii) $\bar{X} = 7$ units $N = 3$ $\sigma_n = 3.266$ units
Put the calculator in statistical mode, clear the memories and enter the data so:
(i) 5 M+ 7 M+ 9 M+
(ii) 3 M+ 7 M+ 11 M+

5 (i) $\bar{X} = 5$ units $N = 3$ $\sigma_n = 1.633$ units
(ii) $\bar{X} = 5$ units $N = 3$ $\sigma_n = 3.266$ units
The distributions have been moved along the variable axis to a different location; however, the spread is exactly the same.

6 (i) 9999 $\bar{X} = 9$ sd = 0
(ii) 0000 $\bar{X} = 0$ sd = 0
(iii) 0099 $\bar{X} = 4.5$ sd = 4.5
(iv) any four identical numbers.

7

x	f	fx	fx^2
1	2	2	2
2	1	2	4
3	4	12	36
4	1	4	16
5	2	10	50
Totals	10	30	108

$$\bar{X} = \frac{\sum fx}{N} = \frac{30}{10} = 3.0 \text{ units}$$

$$\text{variance} = (s)^2 = \frac{\sum fx^2}{N} - \bar{X}^2$$

$$= \frac{108}{10} - 3^2$$

$$= 10.8 - 9 = 1.8$$

Variance = $(s)^2 = 1.8$, and so the standard deviation $s = \sqrt{(1.8)} = 1.34$ units.

8 $\bar{X}$ = 3 children; σ_n = 1.832 children; σ_{n-1} = 1.2472 children.

9

2	15	78
7	16	02234
23	16	5566667788899999
(31)	17	0000011112222223333333344444444
46	17	5555555556666777778899999
20	18	0001111222334
6	18	6688
2	19	23

Median = 174.0 cm; quartiles = 178.75, 170.0.

10

Class	*x*	*f*	*fx*	*fx*²
155–159	157	2	314	49298
160–164	162	5	810	131220
165–169	167	16	2672	446224
170–174	172	31	5332	917104
175–179	177	26	4602	814554
180–184	182	14	2548	463736
185–189	187	4	748	139876
190–194	192	2	384	73728
Totals		100	17410	3035740

$$\text{mean} = \bar{X} = \frac{\sum fx}{N} = \frac{17410}{100} = 174.1 \text{ cms}$$

$$(s)^2 = \frac{\sum fx^2}{N} - \overline{X}^2 = \frac{3035740}{100} - (174.1)^2$$

$$= 30357.4 - 30310.81$$

$$= 46.59$$

variance is 46.59 and so the standard deviation $s = \sqrt{(46.59)} = 6.83$ cm.

11 With the calculator in statistical mode and starting with all registers clear enter the data in the following order:

157 × 2 M+
162 × 5 M+ etc.

to get $N = 100$, $\bar{X} = 174.1$, $\sigma_n = 6.82569$, $\sigma_{n-1} = 6.86007$

12 (i)

```
MTB > READ 'A:\DATA02.DAT' into C1-C3
    100 ROWS READ
ROW       C1       C2       C3
  1        1      173      155
  2        2      181      169
  3        3      169      162
  4        4      183      166
 . . .
MTB > DESCRIBE C2 C3

                N      MEAN    MEDIAN    TRMEAN     STDEV    SEMEAN
HTMEN         100    174.11    174.00    174.06      6.78      0.68
HTWOMEN       100    161.25    162.00    161.40      6.41      0.64
              MIN       MAX        Q1        Q3
HTMEN      157.00    193.00    170.00    178.75
HTWOMEN    143.00    176.00    157.00    165.75
```

(ii)

```
MTB > BOXPLOT C2

                     -----------
    * ---------------I    +    I------------    **
                     -----------
  ------+---------+---------+---------+---------+---->HTMEN
160.0   168.0     176.0     184.0     192.0
```

```
MTB > BOXPLOT C3
                                    -------------
       * *      -------------I     +     I--------------
                                    -------------
--------+---------+---------+---------+---------+--HTWOMEN
      147.0     154.0     161.0     168.0     175.0
```

(iii)
```
MTB > STEM-AND-LEAF C2
Stem-and-leaf of HTMEN      N = 100
Leaf unit = 1.0
     2    15 78
     7    16 02234
    23    16 5566667788899999
  (31)    17 0000011112222223333333344444444
    46    17 5555555556666777777889999 9
    20    18 00011111222334
     6    18 6688
     2    19 23
```

```
MTB > STEM-AND-LEAF C3
Stem-and-leaf of HTWOMEN    N = 100
Leaf unit = 1.0
     1    14 3
     2    14 4
     2    14
     4    14 89
     8    15 0001
    11    15 333
    17    15 444555
    26    15 666667777
    36    15 8888899999
    49    16 0000000011111
  (14)    16 22222223333333
    37    16 444444555555
    25    16 66667777
    17    16 8889999
    10    17 000001
     4    17 22
     2    17 4
     1    17 6
```

By looking at the stem-and-leaf diagrams in conjunction with the boxplots we can see that the outliers for the men are the heights

157, 192 and 193 cm. The outliers for the women are the heights 143 and 144 cm.

From the statistical descriptions we can see that the medians for the men, 174 cm, and for the women, 162 cm, do correspond to the values indicated by the small crosses in the boxplots.

Likewise the ends of the boxes of the two boxplots indicate the quartile values for the two distributions, i.e. for the men 170 and 178.75 cm, and for the women 157 and 165.75 cm.

(iv) The command MTB > HELP BOXPLOT will give the extensive Help program regarding boxplots.

5 Exploratory Data Analysis of Two Data Sets

This chapter is unlike the previous ones in that we will not look at any new methods of data analysis but will focus primarily on the use of the techniques already outlined. Some, however, will be developed or modified. We use Minitab to help us with some of the computation.

Two data sets will be considered, DATA03.DAT and DATA04.DAT. The first is a simple set of heights and weights of students and the second is a set of national suicide statistics abstracted from the World Health Organization (WHO) publication *World Health Statistics Annual 1988*.

The best way to use this chapter is as a guide and commentary to help your analysis. Read the section through quickly and then start your exploratory analysis 'by hand' or in Minitab with the guidance given.

Height Data

Let us look at the first data set (table 5.1). It consists of the height in centimetres and the weight in kilograms, each measured to the nearest integer, of 45 male and 39 female students.

Table 5.1 Heights (cm) and weights (kg) measured to nearest integer

C1	C2	C3	C1	C2	C3	C1	C2	C3
Males $N=45$								
1	158	63	1	175	*	1	180	77
1	159	59	1	175	66	1	180	71
1	164	64	1	176	67	1	181	69
1	166	62	1	176	68	1	182	82
1	166	61	1	176	74	1	182	83
1	167	57	1	176	78	1	183	77
1	169	67	1	177	70	1	185	82
1	170	66	1	177	68	1	185	80
1	170	70	1	177	79	1	186	77
1	171	74	1	178	73	1	188	78
1	171	69	1	178	*	1	189	85
1	172	70	1	178	76	1	190	*
1	173	73	1	178	60	1	194	80
1	174	67	1	179	68	1	196	73
1	174	73	1	180	80	1	199	76
Females $N=39$								
0	148	50	0	160	*	0	167	61
0	150	52	0	161	59	0	168	66
0	151	47	0	161	58	0	170	64
0	154	49	0	161	55	0	173	68
0	154	48	0	162	55	0	174	57
0	155	*	0	163	59	0	174	71
0	155	56	0	164	56	0	181	70
0	156	*	0	164	53	0	182	65
0	157	53	0	164	57	0	185	67
0	157	50	0	164	62			
0	157	*	0	164	56			
0	158	56	0	165	57			
0	159	52	0	165	60			
0	159	53	0	165	61			
0	160	55	0	166	55			

We should note how the data are formatted (see Table 5.1 and Appendix 1). They appear in three columns, the first C1, indicates the sex of the undergraduate (1 = male, 0 = female); the second, C2, gives the height in centimetres, and the third, C3, gives the weight in kilograms. The asterisks denote missing values of the variables.

This first data set is relatively simple and not very large but analysis would be tedious unless we used a computer. So here we dispense with our initial analysis by 'hand' and concentrate on Minitab.

These data are to be found in data file DATA03.DAT and so we will copy it into Minitab with the command

```
MTB > READ 'A:\DATA03.DAT' into C1-C3
```

Note that as this is an ASCII file we use the READ command and we must specify how many columns it is to be read into.

Start saving your analysis by using the OUTFILE command. For example the command MTB > OUTFILE 'A:\GZALA04' will save the analysis into a file in A: called GZALA04.LIS. Such a saved analysis file cannot be called back into Minitab but it can be read, edited and printed out from a word-processing package such as Wordperfect.

The data set contains three variables, sex, height and weight in three columns. Let us suppose that we wish to split the height and weight data into separate male and female columns. This type of operation is common in data analysis. We can go about it as follows using the information in C1 as a control to facilitate the splitting.

Let us split C2 (height) and C3 (weight) into four separate columns as follows: C5 (male height), C6 (female height), C7 (male weight) and C8 (female weight). We need four separate commands to do this:

```
MTB > COPY C2 into C5;
SUBC> USE C1=1. (copies male heights from C2 into C5)

MTB > COPY C2 into C6;
SUBC> USE C1=0. (copies female heights from C2 into C6)
```

```
MTB > COPY C3 into C7;
SUBC> USE C1=1. (copies male weights from C3 into C7)

MTB > COPY C3 into C8;
SUBC> USE C1=0. (copies female weights from C3 into C8)
```

You should pay particular attention to the punctuation in the above commands. The semicolon on the first command line activates the subcommand prompt, and the subcommand line always ends in a full stop. If you make an error in the subcommand line you can get out of it with the command SUBC > ABORT, and then start again.

After giving the four direct commands above go into the data editor and check that you do indeed have 45 male heights in C5, 39 female heights in C6, 45 male weights in C7 and 39 female weights in C8. Go to the space above the top of the columns C5, C6, C7 and C8 in the editor and label them MALEHT, FEMALEHT, MALEWT and FEMALEWT respectively.

Let us now start our analysis by obtaining a simple statistical description with the command, MTB > DESCRIBE C5-C8 we get the following:

	N	N*	MEAN	MEDIAN	TRMEAN
MALEHT	45	0	177.33	177.00	177.27
FEMALEHT	39	0	162.90	162.00	162.51
MALEWT	42	3	71.71	72.00	71.79
FEMALEWT	35	4	57.51	56.00	57.32

	STDEV	SEMEAN	MIN	MAX	Q1	Q3
MALEHT	8.80	1.31	158.00	199.00	171.50	182.00
FEMALEHT	8.43	1.35	148.00	185.00	157.00	166.00
MALEWT	7.07	1.09	57.00	85.00	67.00	77.25
FEMALEWT	6.20	1.05	47.00	71.00	53.00	61.00

N indicates the number of values of the variable in each column, while *N** tells us how many missing values there are.

We should note how close the mean and median are in each of the four distributions above. This is an indication of their symmetry. However, as we would expect, the average (mean and median) male height and weight is greater than the average female height and weight.

Let us now examine the four distributions using the histogram and stem-and-leaf facility. That is, we give the commands

```
MTB > HISTOGRAM C5 C6 C7 C8
MTB > STEM-AND-LEAF C5 C6 C7 C8
```

Look at the distributions shown on the screen and you will see that as we surmised the four distributions do seem to have a reasonable degree of symmetry about their averages, but the male heights are noticeably more symmetrical than the others. On the other hand male weights are the least symmetrical and are almost bimodal.

Let us now use the simplest type of plot instruction in Minitab to draw a graph of, say, male height in C5 against male weight in C7:

```
MTB > PLOT C5 against C7
```

A similar plot for females can be obtained with the command

```
MTB > PLOT C6 against C8
```

We note that these plots give us a lozenge-shaped scatter of points going from lower left to upper right of the graph, indicating what we already know – that taller people generally weigh more than short people. We will return to such plots in chapter 6 when we look at the correlation of variables.

We can obtain individual boxplots of the distributions by using commands such as

```
MTB > BOXPLOT C5
MTB > BOXPLOT C6
MTB > BOXPLOT C7
MTB > BOXPLOT C8
```

However, this would result in separate plots and it would be more convenient for comparison to have all the height plots on one diagram and all the weight plots on another. This is most easily achieved using the BY command and the height and weight data in an unsplit form, i.e. C2 and C3. Try the commands

```
MTB > BOXPLOT C2;
SUBC> BY C1.

MTB > BOXPLOT C3;
SUBC> BY C1.
```

The first command gives a boxplot of male and female heights on the same axis, while the second gives a boxplot of male and female weights on the same axis. Again note the punctuation.

The first pair of boxplots, those of female and male heights, appear as in figure 5.1.

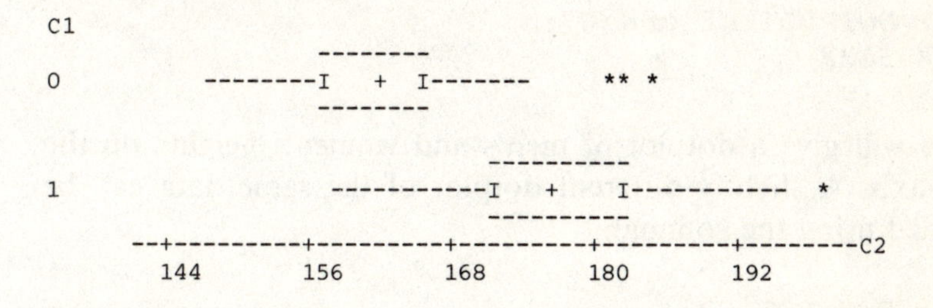

```
C1

                   --------
0         --------I   +  I-------        ** *
                   --------

                             ----------
1                ----------I    +    I-----------   *
                             ----------

   --+---------+---------+---------+---------+--------C2
    144       156       168       180       192
```

Figure 5.1

As we noted from the statistical description earlier, the average height of the women, as measured by the median (small plus in the boxes), is less than that of the men, but the spread of the two distributions as indicated by the midspread (size of the boxes) is very similar. Both distributions have tall outliers, with three females being considerably taller than the average male.

We should note at this point that it is often necessary to split data into separate columns as we did at the beginning of this exercise by using a control column (in our case the control column was C1 which contained the control symbols 1 and 0, i.e. male and female). However, we do not always need to do this as Minitab, with the BY and other commands, can give separate plots on the same axis using a control column. In data analysis it is sometimes necessary to split data into separate columns but we should bear in mind what the various subcommands can do. It is worth noting in this context that although we only had two values, 0 and 1, in our control column C1, Minitab can utilize more control codes.

Another useful graphic comparison can be obtained using the DOTPLOT function. The dotplot gives a distribution which is similar to the histogram but the axis is divided into more divisions and data values are represented by a dot.

Try a simple dotplot first. MTB > DOTPLOT C5. This command will give you a dotplot of the 45 male heights that we put into column C5. You should now give the boxplot command MTB > BOXPLOT C5 and compare the two plots and remind yourself of what the various parts of the boxplot indicate. Check in the last chapter if you have forgotten.

The subcommand SAME is also useful and can be used with a variety of commands. For instance try the following:

```
MTB > DOTPLOT C5 and C7;
SUBC> SAME.
```

This will give a dotplot of men's and women's heights on the same axis. A slightly different dotplot of the same data can be obtained using the command

```
MTB > DOTPLOT C2;
SUBC> BY C1.
```

```
                           :
C1                         :
0                    .. .  :
             . .. ::.::::..:... . .:     .. .
         -+---------+---------+---------+---------+---------+C2

                                     .
C1                                   :
1                               :    :: . .
                      :    .:. .:..::::.:.: :. :.   .. .
         -+---------+---------+---------+---------+---------+C2
       144       156       168       180       192       204
```

Figure 5.2

This command gives us the plots shown in figure 5.2, which give a different picture of the two distributions of height that we examined previously using boxplots. It is often instructive to use boxplots and dotplots in conjunction to look at a distribution. We should note that in the above we can clearly 'see' the outliers that our earlier boxplot alerted us to.

As you probably realize, each of the commands, such as HISTOGRAM, BOXPLOT etc., has various subcommands of which we have only looked at two, i.e. BY and SAME. For details of the various subcommands you should consult the Minitab manual or ask for HELP SUBCOMMANDS while in Minitab. These various subcommands are also reasonably clearly laid out as options in the Menu Command Dialogue Boxes.

We will finish our brief exploratory analysis of the heights and weights here, but before we do I leave you with a small puzzle. Use the command

```
MTB > LPLOT data in C2 against data in C3
      using tag values in C1
```

Try to work out what this plot tells us. (*Hint*. You may like to compare it with MTB > PLOT C2 against C3.) What do you think the As and the Zs represent? A few 2s are also to be seen, what do they signify?

This brings us to the end of our first exploratory analysis of our simple data set of heights and weights. Please pause before starting the analysis of the next data set and write down a brief résumé of what you have found from your analysis.

Suicide Data

Let us now look at a larger and slightly more complex data set and see what sense we can make of it. The data consist of the suicide rates from 16 countries selected from the WHO publication *World Health Statistics 1988*.

When faced with a complex data set such as this the first approach to analysis should be by hand. Certain parts of the data can easily be analysed in this fashion and will suggest approaches for later computer analysis.

I will say a few words later to explain why I chose these particular statistics, but first I would like you to look carefully at the data (table 5.2) and then jot down the points which strike you about the form and distribution of the statistics. Please do this before you read what I have written.

A variety of things come to mind as we first glance at the data in the table. The rates are for various countries and are classified by

Table 5.2 Age-specific death rates per 100,000 population for suicide

Country	Sex C1	All C2	15–24 C3	25–34 C4	35–44 C5	45–54 C6	55–64 C7	65–74 C8	75+ C9
Austria	1	40.1	29.3	42.9	44.8	54.9	48.5	68.1	125.2
	0	15.7	8.1	10.7	15.9	16.8	22.9	33.8	35.4
France	1	32.9	16.0	34.2	38.1	46.3	47.8	63.5	121.3
	0	12.9	4.6	10.1	14.4	20.3	21.2	25.4	29.5
Germany	1	26.7	17.6	24.3	27.7	35.2	37.5	43.9	77.2
	0	11.8	4.5	7.8	10.4	15.6	17.4	23.2	23.7
Hungary	1	65.9	24.3	67.3	89.3	111.5	96.5	118.4	188.0
	0	25.6	10.3	17.9	25.8	33.7	37.3	56.4	69.0
Ireland	1	12.1	14.0	18.7	13.1	17.7	28.3	20.5	8.8
	0	3.8	3.3	5.2	5.3	9.2	8.3	4.7	1.1
Israel	1	8.5	6.6	7.7	9.8	13.9	19.7	26.7	39.8
	0	4.5	4.5	3.7	5.6	4.6	12.6	8.8	17.5
Italy	1	12.2	5.2	9.2	10.1	14.8	22.0	33.7	50.0
	0	4.7	1.3	2.7	4.3	5.9	8.8	11.8	12.4
Netherlands	1	13.9	8.1	15.2	15.3	19.6	22.6	25.5	42.8
	0	8.2	3.6	9.1	10.9	13.6	13.5	12.5	13.0
Poland	1	22.3	17.8	30.6	32.3	38.5	35.0	29.9	33.2
	0	4.7	3.0	5.7	6.4	8.0	7.7	7.7	6.9
Spain	1	9.8	5.9	7.9	9.3	15.1	19.0	22.2	38.4
	0	3.4	1.6	2.2	2.5	4.8	6.5	7.9	10.9
Sweden	1	27.1	19.5	29.4	38.8	38.8	36.1	39.3	48.0
	0	10.1	7.9	11.0	14.6	15.4	15.3	12.7	8.8
Switzerland	1	35.0	27.7	37.3	41.3	44.5	43.3	56.7	87.3
	0	13.7	8.2	11.8	13.8	15.1	18.7	30.0	27.4
England and Wales	1	11.6	9.3	14.6	14.6	15.0	17.8	15.8	20.0
	0	4.5	2.1	3.8	4.6	7.6	7.3	8.1	6.9
Canada	1	22.8	26.9	32.0	28.5	28.1	27.6	28.4	36.3
	0	6.4	5.3	7.5	8.9	11.4	8.9	9.2	5.5
USA	1	20.6	21.7	25.5	23.0	24.4	26.7	35.5	56.0
	0	5.4	4.4	5.9	7.6	8.8	8.4	7.3	6.8
Japan	1	25.6	11.6	23.7	29.4	45.5	40.5	42.1	73.0
	0	13.8	6.5	9.9	10.9	16.9	21.2	31.1	53.2

1, male; 0 female. C2–C9 are age bands in years.

Source: World Health Organization, 1988

sex and age cohorts. The mix of countries might seem odd – most are European, all are industrial and capitalist – but why this particular 16?

The choice was dictated by precedent. The countries, with the exception of Ireland, were chosen by Erikson and Nosanchuk in their book *Understanding Data*. They use similar WHO figures for 1971 and I thought it would be instructive to rework some of the analysis with more modern data so that students may, if they wish, compare trends over time. You will find that I have not exactly duplicated their analysis and I have abstracted more age cohorts than they used. Also as mentioned, I have included Ireland which shows a rather contrary and interesting trend.

As you are probably aware the study of suicide and the comparison of national suicide rates has occupied an important place in social science since Emile Durkheim's classic study *Suicide* was published in 1897. Our focus on this course is the analysis of data but it is well worth keeping Durkheim's contentions in mind. His main argument is that suicide, rather than being looked at as a personal problem related to an individual's characteristics and particularly any mental illness (... he took his life whilst the balance of his mind was disturbed ...), should be related to the organisation of a society, that is, to its structure. He claims that comparative statistics show that suicide is best understood as a social rather than an individual phenomenon. It should be related to social solidarity and integration rather than the individual states of mind of the men and women who kill themselves.

Maybe the choice of countries did not strike you as odd but you will certainly have noticed the great difference between male and female rates. In each country the female rate is about a half or less than the male rate. Again, reading across the age cohorts you will have noticed that most countries show a rise of suicide rate with age for both males and females. In some cases this rise with age is steep, in others more gradual. The Irish figures are at odds with this trend.

You will also have noticed the disparity in rates between the various countries and probably noted the Hungarian rate. It has the highest general rate for both men and women. The Hungarian rate for the first age cohort of 15–24 years is high for each sex and it rises to dramatic proportions for the last age cohort. What might we imagine Durkheim to say about the social integration of ageing Hungarians?

After these first impressions let us start our analysis. Let us first plot the overall country rates to the nearest unit for each sex using a back-to-back stem-and-leaf distribution.

Male Leaf	Stem	Leaf Female
6	6	
	5	
0	4	
53	3	
776321	2	6
42220	1	023446
9	0	345555568

Stem: tens Leaf: units $N = 16$

The male and female distributions are each centred about a different level (average) and the male level is considerably higher than the female. The spread of data also differs with the females having the smaller spread. The female data are skewed in that they are concentrated at the low end of the scale.

It is useful to remind ourselves of which general rates belong to which country so that we can twin the above diagram with one showing country leaves.

Males Leaf	Stem	Females
HU	6	
	5	
AU	4	
SZ.FR	3	
SW.GR.JA.CA.PO.US	2	HU
NE.EW.IT.IR.SP	1	SW.GR.FR.SZ.JA.AU
IS	0	SP.IR.IS.IT.PO.EW.US.CA.NE

stem: tens leaf: countries N = 16

These stem-and-leaf distributions of the countries might suggest a straightforward ranking of the overall rates. The country with the highest suicide rate is ranked 16, the country with the next highest rate 15, and so on until we get to the country with the lowest rate which is ranked 1. These ranks are found by counting from either end of the stem-and-leaf distributions.

	Male	*Female*
Hungary	16	16
Austria	15	15
Switzerland	14	13.5
France	13	12
Sweden	11.5	10
Germany	11.5	11
Japan	10	13.5
Canada	9	8
Poland	8	5
USA	7	5
Netherlands	6	9
Italy	4	5
Ireland	4	2
England and Wales	4	5
Spain	2	1
Israel	1	5

Countries with equal rates are given equal ranks. For example, the male rate for Sweden and Germany is the same so they both have the rank $1/2\ (11+12) = 11.5$. Similarly males in England and Wales, Ireland and Italy have rank $1/3\ (3+4+5) = 4$.

There does appear to be a fairly high degree of correlation between the male and female ranks but whether it is meaningful or significant remains to be seen. We will look at ideas of correlation in a later chapter. Here we shall merely note that if the correlation is significant this seems to indicate that structural or societal forces are important but that they have a differential effect on males and females in any one society.

From what we have seen so far it is obvious that country and sex are important variables in determining the suicide rate. A glance across the age cohorts indicates that age is another important variable. An analysis of all the age groups can be left to the computer, but we can look first at the extreme ends of the age spectrum. We will plot the figures for young and old males into stem-and-leaf distributions and then illustrate them by boxplots.

Males 15-24 years		Males 75+ years
Leaf	Stem	Leaf
	18	8
	.	
	.	
	.	
	12	15
	11	
	10	
	9	
	8	7
	7	37
	6	
	5	06
	4	038
	3	368
987420	2	0
88642	1	
98765	0	9
N = 16	stem: ten	leaf: units

From the above distributions by counting and calculation we can obtain the following values:

	15–24 years	*75+ years*
N	16	16
x_u	29	188
x_{qu}	23.5	84.5
x_m	17	49
x_{ql}	8.4	36.5
x_l	5	9
d_q	15.1	48
$\bar{X}$	16.3	65.3
s	8.1	46.2

From a consideration of these two extreme age cohorts it seems that both the level and the spread of the rates increase with age. This will be investigated later with computer-generated boxplots for all age groups. Here we should note that we can get a visual comparison of the spread of the two distributions by subtracting the median from each and replotting the resulting boxes and whiskers.

	15–24 years	*75+ years*
N	16	16
x_u	12	139
x_{qu}	6.5	35.5
x_m	0	0
x_{ql}	−8.6	−12.5
x_l	−12	−40

By removing the level, in this case subtracting the median from the values, other features of the distribution such as the spread become more apparent.

We may go one step further and in addition to removing levels in this way we can also remove spread if we wish. This is called standardization and always takes the form

$$\text{standardized score} = \frac{\text{value} - \text{level}}{\text{spread}}$$

We may standardize by using the mean and standard deviation to get

$$\text{standardized score} = \frac{x - \bar{X}}{\text{standard deviation}}$$

This particular standardized score using the mean and standard deviation is commonly called the Z score, and we shall come across it in later chapters.

We may also standardize using the median and the midspread d_q to get

$$\text{standardized score} = \frac{x - \text{median}}{d_q}$$

Once data are standardized in this fashion they will have a zero level (average) as in the cases plotted above and each distribution will have a spread of one unit. Standardization allows us to compare distributions more easily, and once both level and spread have been removed other characteristics of the distribution such as skew (lopsidedness) may be seen more easily.

Let us calculate just two standardized scores using the median and midspread for a couple of the values of the extreme age cohorts looked at earlier, that is, x_{qu} for the group 15–24 years old and x_{ql} for the group 75+ years.

15-24 years

$$\text{standardized } x_{qu} = \frac{\text{value} - \text{level}}{\text{spread}} = \frac{x_{qu} - x_m}{d_q}$$

$$= \frac{23.5 - 17}{15.1} = 0.43$$

75+ years

$$\text{standardized } x_{ql} = \frac{\text{value} - \text{level}}{\text{spread}} = \frac{x_{ql} - x_m}{d_q}$$

$$= \frac{36.5 - 49}{48} = -0.26$$

In a similar fashion we can standardize the other values using the median and midspread to get,

	15–24 years	*75+ years*
N	16	16
x_u	0.79	2.90
x_{qu}	0.43	0.74
x_m	0	0
x_{ql}	–0.57	–0.26
x_l	–0.79	–0.83

The three boxplots in figures 5.3–5.5 are shown one above the other for comparison. The first clearly shows the difference in level between the two distributions and the second, with the medians removed, enables us to appreciate the difference in spread between the distributions. The third, the standarized plot, has both level and spread removed and gives an indication of the skewness or lopsided nature of each distribution.

As mentioned above, standardization by the removal of the level and spread from data allows us to compare distributions more easily. However, a few words of caution are necessary regarding the word 'removal'. It is not to be thought of as removal in the sense of eliminating them or removing them from consideration. Tukey calls it 'setting aside' and what we are doing is temporarily setting aside the level and spread so that we can examine the

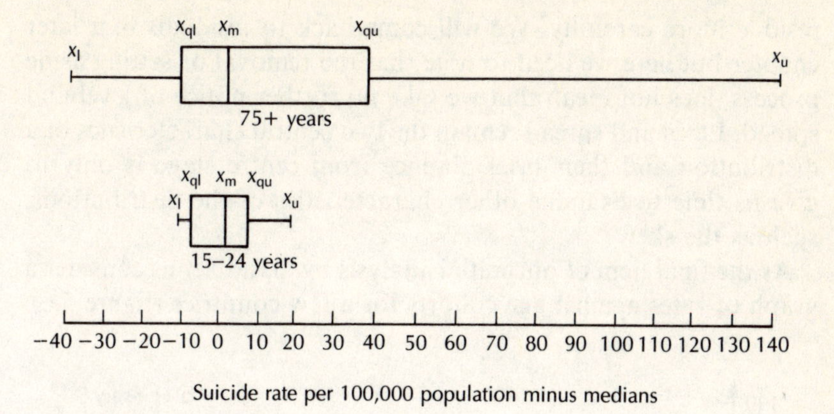

Figure 5.3

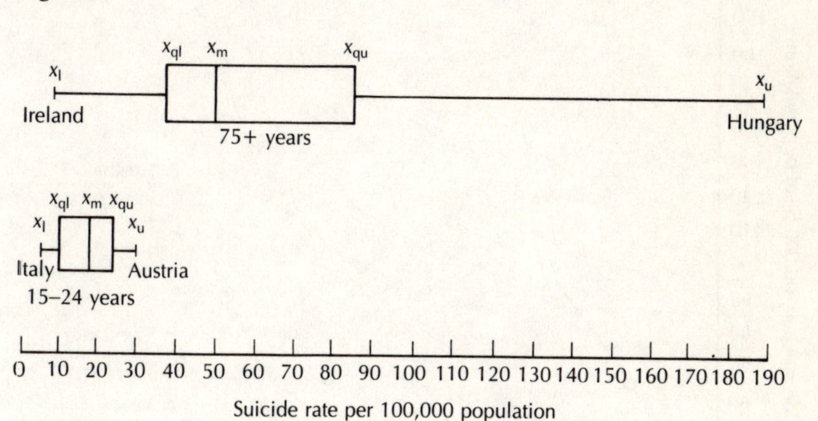

Figure 5.4

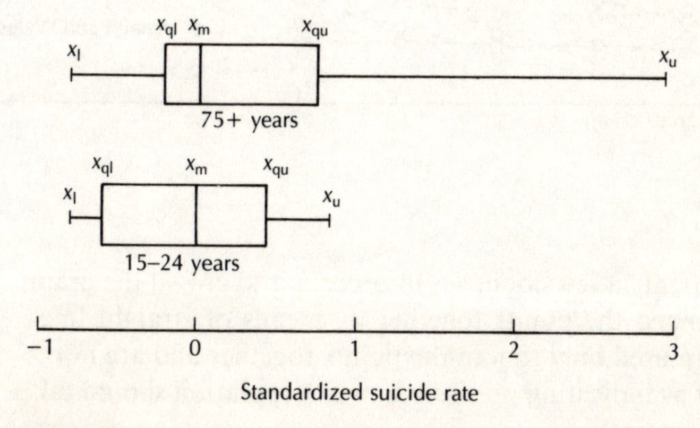

Figure 5.5

residue more carefully. We will come back to residuals in a later chapter but here we need to note that the removal or setting aside process does not mean that we take no further notice of level and spread. Level and spread remain the two central characteristics of a distribution and their brief absence from centre stage is only to give us time to examine other characteristics of the distributions, such as the skew.

As the final item of our initial analysis by 'hand' let us consider a graph of rates against age cohorts for a few countries (figure 5.6).

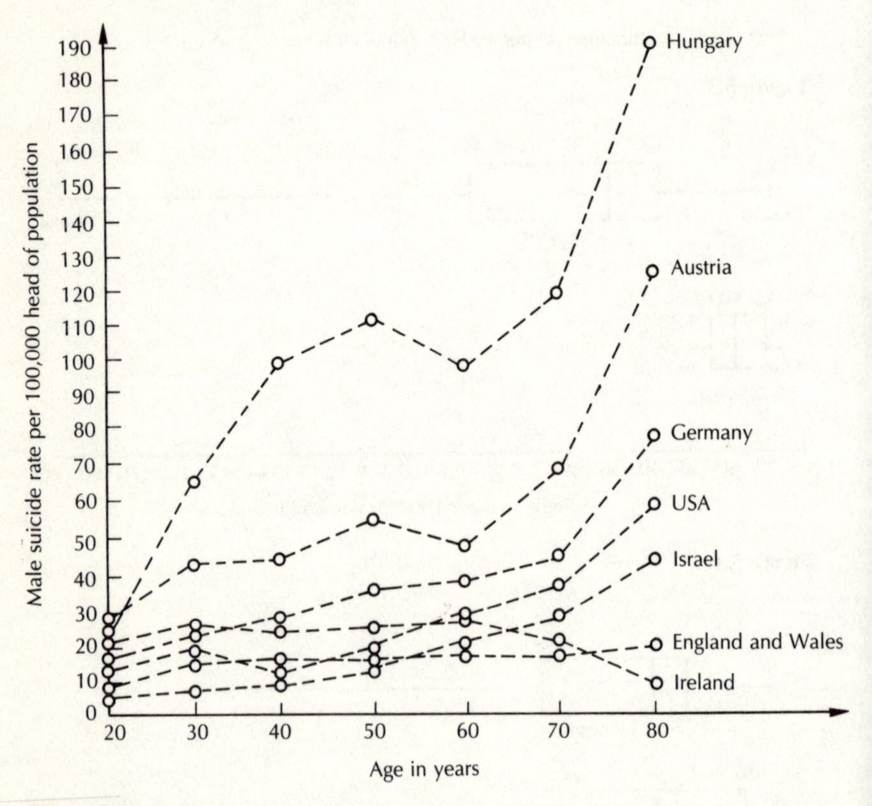

Figure 5.6

I have chosen only a few countries in order not to crowd the graph and I have joined the points together by means of straight lines. These lines are used only to join the points together and are not to be thought of as indicating precisely how interpolation should take place between points.

Although we have only plotted a few of the countries some trends do seem evident and worthy of further study. Generally rates start relatively low and increase fairly steadily until early middle age; for many there seems then to be a slackening of the rate of increase and for some even a drop in the rate. However, after this the rate climbs steeply with age. The steepness of the last portion of the trace (70–80 years) might be misleading, however as I have arbitrarily plotted the figures for 75+ years at 80 years.

Many countries, e.g. Israel, the Netherlands (not shown), and Italy (not shown) have similar traces and rates. Some countries have similarly shaped traces but are based on different levels. Canada (not shown) and England and Wales have similar traces but Canada is much higher. Ireland is the only country to have a peak rate at 60 years and then a decline with old age.

Before going on to enlist the aid of the computer in our analysis a few brief comments on the limits of what we can get from a set of data may be appropriate. Learning to look at and into data is a craft worth learning and it is made easier by these exploratory data analysis techniques developed by Tukey and others. With data sets of any complexity we need to start analysis by 'hand' but once we have the feel of the data the computer can help with repeated or tedious analysis. However, despite the techniques and computing power now at our disposal, we should not imagine that we can find out everything about a data set merely by analysing it assiduously. While we may find patterns in the data or in other cases impose or fit patterns to them we should be aware of the limitations of our methods. The techniques we are using are properly called exploratory and there are many characteristics of data sets which cannot be understood from a close study of the data themselves. Many characteristics or patterns are inexplicable unless we also consider factors outside the data set. Such a consideration is beyond the scope of this text but a few words of caution are in order.

Perhaps an analogy from social science may help. If we wished to study what goes on in, say, a hospital or factory we would quite rightly focus our attention on factors within those institutions. We would study the form of the division of labour in the organization, how the groups of workers relate to each other, the role of technology, who has power and authority and how the rewards of work are distributed etc. However, not all the meaning of what goes on in such complex organizations can be found by looking at intra-organizational factors. Such organizations are set within different types of modern societies and states whose cultures, values and practices need to be taken into account to make sense of some of the patterns found within the hospital or factory. For instance, social divisions of class, race, ethnicity or gender in the larger society may be crucial to an understanding of who does what in such organizations in that society.

Similarly the close study of a data set can tell us a great deal but not all. Some patterns or characteristics may be inexplicable unless we look outside the data set. We may need to ask, for instance, how the data were gathered, for what purposes and under what conditions. What were the cultural, political or economic constraints on collection? Are such data homogeneous or consistent enough to be used for secondary analysis? For example, we could ask ourselves how reliable the suicide data for Ireland is. It seems to show a level and a trend which is somewhat odd. This may be a reflection of the incidence of suicide and the role of religion in societal integration or it could be related to, say, medical or other decisions and to the collection of statistics. Is there perhaps an unwillingness of doctors in a strongly Catholic country to ascribe death to suicide unless they clearly cannot avoid it? A consideration of such questions and the existence of possibly powerful external factors should always condition our analysis of data.

After our preliminary analysis, let us now continue with our exploration of the data using the computer to help us. The file of this data set of the suicide statistics is DATA04.DAT. You should copy it from the floppy disk with the command MTB > READ 'A:\DATA04.DAT' into C1-C9. Remember that this is an ASCII file and that with the command READ we have to specify the number of columns the data are to be read into.

Start to save your analysis by giving the OUTFILE command, e.g. MTB > OUTFILE 'A:\SUICIDE2'. This will save the analysis in a file in A: called SUICIDE2.LIS. Later we shall also save our worksheet with the SAVE command.

In this analysis we look primarily at the male data for suicide and I leave it for you to do a parallel analysis of the female data should you so wish.

Let us take the male data out of columns C2–C9 using the COPY command that we used earlier with the height and weight data. The command below will take the male data out of C2 and put it into C12.

```
MTB > COPY C2 into C12;
SUBC> USE C1=1.
```

The command to take the male data out of C3 and put it into C13 is

```
MTB > COPY C3 into C13;
SUBC> USE C1=1.
```

Using six similar commands, you should now take the male data out of the remaining columns C4, C5, C6, C7, C8 and C9 and copy it into columns C14, C15, C16, C17, C18 and C19 respectively. Now go into the editor and check that you have the male data in columns C12–C19. Label the columns as follows: MALEALL, M15-24, M25-34 etc. Ask for information with the command MTB > INFO, and we will print our male data into the analysis file with the command, MTB > PRINT C12-C19.

```
COLUMN     NAME        COUNT
C1                        32
C2                        32
C3                        32
C4                        32
C5                        32
C6                        32
C7                        32
C8                        32
C9                        32
C12        MALEALL        16
C13        M15-24         16
C14        M25-34         16
C15        M35-44         16
C16        M45-54         16
C17        M55-64         16
C18        M65-74         16
C19        M75+           16
CONSTANTS USED: NONE
```

ROW	MALEALL	M15-24	M25-34	M35-44	M45-54	M55-64	M65-74	M75+
1	40.1	29.3	42.9	44.8	54.9	48.5	68.1	125.2
2	32.9	16.0	34.2	38.1	46.3	47.8	63.5	121.3
3	26.7	17.6	24.3	27.7	35.2	37.5	43.9	77.2
4	65.9	24.3	67.3	89.3	111.5	96.5	118.4	188.0
5	12.1	14.0	18.7	13.1	17.7	28.3	20.5	8.8
6	8.5	6.6	7.7	9.8	13.9	19.7	26.4	39.8
7	12.2	5.2	9.2	10.1	14.8	22.0	33.7	50.0
8	13.9	8.1	15.2	15.3	19.6	22.6	25.5	42.8
9	22.3	17.8	30.6	32.3	38.5	35.0	29.9	33.2
10	9.8	5.9	7.9	9.3	15.1	19.0	22.2	38.4
11	27.1	19.5	29.4	33.8	38.8	36.1	39.3	48.0
12	35.0	27.7	37.3	41.3	44.5	43.3	56.7	87.3
13	11.6	9.3	14.6	14.6	15.0	17.8	15.8	20.0
14	22.8	26.9	32.0	28.5	28.1	27.6	28.4	36.3
15	20.6	21.7	25.5	23.0	24.4	26.7	35.5	56.0
16	25.6	11.6	23.7	29.4	45.5	40.5	42.1	73.0

To start our analysis let us ask for a statistical description of the data in these columns with the command MTB > DESCRIBE C12-C19.

	N	MEAN	MEDIAN	TRMEAN	STDEV	SEMEAN
MALEALL	16	24.19	22.55	22.34	14.67	3.67
M15-24	16	16.34	16.80	16.21	8.12	2.03
M25-34	16	26.28	24.90	24.68	15.29	3.82
M35-44	16	28.77	28.10	25.84	19.92	4.98
M45-54	16	35.24	31.65	31.31	24.42	6.11
M55-64	16	35.56	31.65	32.47	19.16	4.79
M65-74	16	41.87	34.60	38.26	25.44	6.36
M75+	16	65.3	49.0	60.6	46.2	11.6

	MIN	MAX	Q1	Q3
MALEALL	8.50	65.90	12.13	31.45
M15-24	5.20	29.30	8.40	23.65
M25-34	7.70	67.30	14.75	33.65
M35-44	9.30	89.30	13.48	37.02
M45-54	13.90	111.50	15.75	45.25
M55-64	17.80	96.50	22.15	42.60
M65-74	15.80	118.40	25.73	53.50
M75+	8.8	188.0	36.8	84.8

We could also form the stem-and-leaf plots with the command MTB > STEM-AND-LEAF C12-C19.

In our earlier analysis by hand we formed boxplots of the two extreme age cohorts of 15–24 years and 75+ years on the same axis. It would be useful to get Minitab to boxplot all the age cohorts in a similar fashion. However, first we need to manipulate the data as follows.

First we stack all the data in columns C12–C19 on top of one another in one column, say C30. Each of the eight columns C12–C19 contains 16 pieces of data and so column C30 now contains $8 \times 16 = 128$ rows or pieces of data. We now need another column, say C31, with the same number of rows to plot against it. We get this by putting 16 twos, 16 threes, 16 fours etc. into C31 with the SET command. Columns C30 and C31 should now each contain 128 rows and we can plot one against the other. The commands are as follows:

```
MTB > STACK C12-C19 and put into C30
MTB > SET C31
DATA> 16(2) 16(3) 16(4) 16(5) 16(6) 16(7) 16(8) 16(9)
DATA> END
```

After stacking C30 with the 128 pieces of data and setting C31 with an equal number of control symbols 2, 3, 4, ..., 8, 9 with the above commands, look at the data in the editor using the arrows or page-down key to examine C30 and C31.

Now boxplot C30 against the eight sets of control figures in C31 with the command

```
MTB > BOXPLOT C30;
SUBC> BY C31.
```

This should result in boxplots on the same axis as in figure 5.7.

As we can see from the boxplot for men, the average rate of suicide increases with age. This is confirmed when we examine the changes in the mean and median in the statistical description. There is some flattening of this rate of increase from about 45 years until the early 60s.

As was suggested by our initial boxplots by 'hand', the spread of the data in each age group, as measured by the standard deviation

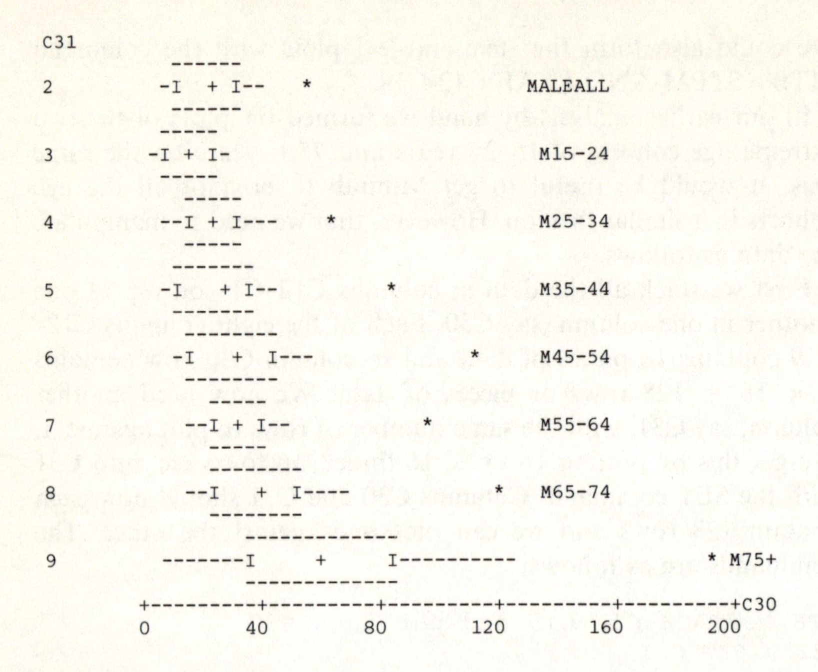

Figure 5.7

and the midspread, also increases with age. Here, however, the rate of increase is less even.

A discontinuity seems to occur with the oldest group of men: both the mean and the median are high and indicate a very heavy suicide rate amongst this group. We should be aware, however, that part of this sudden increase may be due to the conflation of all age categories of 75 years and over. We note also that for this group the mean and the median have become widely separated indicating a very skewed distribution of data. This can also be seen from the stem-and-leaf diagram and the boxplot. That is, we note that in the boxplot for the oldest group of men, the median in the box is placed well over to the left and there is a long tail or whisker to the right.

We have now put a considerable amount of work into this analysis and this is probably a good point for a break. After our earlier OUTFILE command our analysis will have been saved into a file in A: called SUICIDE2.LIS. We shall now end the OUTFILING process with the command MTB > NOOUTFILE.

As we have also considerably modified the original data file it is well worth saving so that we can continue to work on the data later. Give the command MTB > SAVE 'A:\SUICIDE2' and this will save the modified data file into A: in a file called SUICIDE2.MTW. We now end the Minitab session with MTB > STOP.

After a break let us continue our analysis of the suicide data using the saved data file. Go into Minitab and retrieve the file with the command

MTB > RETRIEVE 'A:\SUICIDE2.MTW'. This is a Minitab file rather than an ASCII one so we don't need to specify the columns into which to put it.

We will now save our analysis with the OUTFILE command MTB > OUTFILE 'A:\SUICIDE3' which will give us a file in A: called SUICIDE3.LIS. In conjunction with this OUTFILE command we can also use a command mentioned in chapter 2 but which we have not used so far – a command to limit the width of the output from Minitab. Give the command MTB > OW = 70; this sets the output width of the OUTFILE to 70 characters. It is useful to set the output width in this way so that we can deal adequately with the file in a word-processing package such as Wordperfect where the screen width of 80 can also be set to 70, i.e. with margins on each side of 5 characters. Failure to set the width of Minitab output may result in the wraparound of graphs and tables in the word-processing package.

After having retrieved the data into Minitab it is probably worth looking at it in the editor to remind ourselves of the various columns and also using the command MTB > INFO to give

```
COLUMN     NAME       COUNT
C1                       32
C2                       32
C3                       32
C4                       32
C5                       32
C6                       32
C7                       32
C8                       32
C9                       32
```

```
C12        MALEALL       16
C13        M15-24        16
C14        M25-34        16
C15        M35-44        16
C16        M45-54        16
C17        M55-64        16
C18        M65-74        16
C19        M75+          16
C30                     128
C31                     128
```

Let us now continue our analysis by comparing the overall suicide rate for women and men. We first separate the female data in C2 and put it into C11 with the command

```
MTB > COPY C2 into C11;
SUBC> USE C1=0.
```

Now go into the editor and label the overall female rates in C11 as FEMALALL. Print the overall male and female rates into the analysis file with the command MTB > PRINT C11 C12. This gives us the following:

```
ROW   FEMALALL   MALEALL

  1       15.7      40.1
  2       12.9      32.9
  3       11.8      26.7
  4       25.6      65.9
  5        3.8      12.1
  6        4.5       8.5
  7        4.7      12.2
  8        8.2      13.9
  9        4.7      22.3
 10        3.4       9.8
 11       10.1      27.1
 12       13.7      35.0
 13        4.5      11.6
 14        6.4      22.8
 15        5.4      20.6
 16       13.8      25.6
```

At this point it is worth reminding ourselves of something that we noticed about the male and female rates of suicide when we first looked at the figures before starting our analysis. We noticed the obvious fact that the male rates for all countries were very much higher than the female rates and we also noted that countries that had high male rates also had relatively high female rates. This is probably something to do with what Emile Durkheim would call the social structure in each country and how men and women are integrated into the society.

Previously we ranked the 16 countries by both their male and female overall suicide rates by hand; let us now use Minitab for this task. To find the rankings we use the following commands and then print the rankings into the analysis file.

```
MTB > RANK C11 and put values in C21
MTB > RANK C12 and put values in C22
MTB > PRINT C21 C22
```

C21 contains the female rankings and C22 the male.

ROW	C21 Female	C22 Male	
1	15.0	15	Austria
2	12.0	13	France
3	11.0	11	Germany
4	16.0	16	Hungary
5	2.0	4	Ireland
6	3.5	1	Israel
7	5.5	5	Italy
8	9.0	6	Netherlands
9	5.5	8	Poland
10	1.0	2	Spain
11	10.0	12	Sweden
12	13.0	14	Switzerland
13	3.5	3	England and Wales
14	8.0	9	Canada
15	7.0	7	USA
16	14.0	10	Japan

These Minitab rankings look different to our previous rankings by hand, there are less tied values as the data have not been rounded. Are the male and female rankings for each country similar? An examination of the rankings does indeed show us that countries which have high male rates also tend to have high female rates. For instance the country in row 4, Hungary, is ranked highest at 16 for both male and female rates. Austria, in row 1, is ranked the next highest at 15 for both males and females. The lowest male ranking of 1 occurs in Israel (Durkheim noted that countries at war have a low registered suicide rate). The lowest ranking for the overall female rates occurs in Roman Catholic Spain.

The half values in the female rankings indicate tied values. For instance both Israel and England and Wales have a female overall rate of 4.5 and so both are given the same rank of 3.5.

We might ask ourselves at this stage how close these rankings are. Is there some correlation between the male and female rates? We can get some idea of this relationship by giving the command MTB > PLOT C11 against C12. This gives us the graph in figure 5.8.

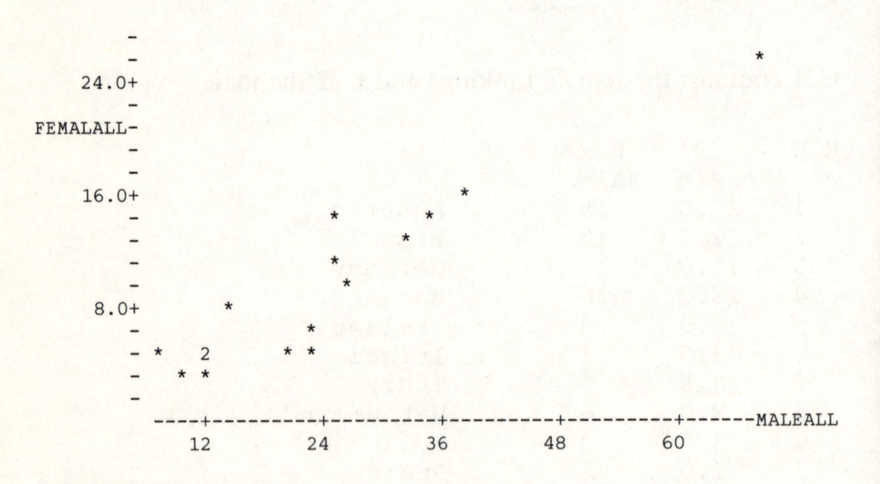

Figure 5.8

The pattern in the displayed data is more or less linear, i.e. we could draw a straight line from bottom left to upper right of the graph which would indicate the approximate pattern of the data points. We look at this type of relationship between two variables in the next chapter on correlation.

Although it probably will not mean much to you if you have not yet studied correlation it is worth asking Minitab to correlate the male and female rates using the command

MTB > CORRELATE C11 and C12. Minitab will then tell you that the correlation of FEMALALL and MALEALL is 0.947. As we shall see in the next chapter, correlation coefficients like this vary between the limits of -1 and $+1$. We will also find that a correlation of 0.947 for 16 data pairs indicates a very high correlation indeed.

Let us return for a moment to the graph of the male rate against the female rate i.e. Fig 5.8. It would be instructive to know what countries the asterisks on the graph represented. We can do this by assigning tag values or labels to each country as follows. Go into the data editor and, starting from the top, type and enter the numbers 1, 2, 3 ,..., 16 in the first 16 rows of column C10. Label C10 TAGS.

These numbers will be used by Minitab to give alphabetical tags to the various countries. For example 1 = A stands for Austria, 2 = B stands for France, 3 = C stands for Germany and so on. The 16 numbers and their equivalent letters are given to the various countries in the order listed in the ranking table above.

The tagged graph is plotted using the LPLOT command which I asked you to investigate earlier. Here the appropriate command is MTB > LPLOT C11 by C12 using the tag values in C10. This gives us the graph in figure 5.9.

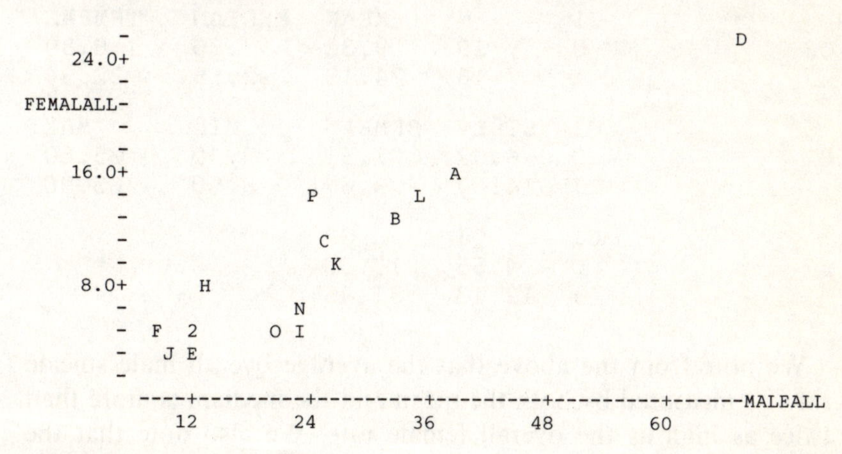

Figure 5.9

As can be clearly seen, country D, i.e. Hungary, has the highest rate for both male and female suicides.

Let us now compare the overall male and female rates by looking at the stem-and-leaf diagrams with the command

```
MTB > STEM-AND-LEAF C11 C12
```

Finally we end our comparison of the overall rates by looking at the boxplots and statistical descriptions with the commands

```
MTB > BOXPLOT C2;
SUBC> BY C1.

MTB > DESCRIBE C2;
SUBC> BY C1.
```

This will give you the following information.

```
C1
                  -------
0               -I  +  I--------
                  -------

                        -------------
1                     --I     +     I------- *
                        -------------

            +---------+---------+---------+---------+---------+C2
            0        15        30        45        60        75

                 C1        N     MEAN   MEDIAN    TRMEAN
C2                0       16     9.32     7.30      8.59
                  1       16    24.19    22.55     22.34

                 C1    STDEV   SEMEAN      MIN       MAX
C2                0     6.02     1.51     3.40     25.60
                  1    14.67     3.67     8.50     65.90

                 C1       Q1       Q3
C2                0     4.55    13.50
                  1    12.13    31.45
```

We note from the above that the average overall male suicide rate, as measured by both the mean and the median, is more than twice as high as the overall female rate. We also note that the female data are more tightly packed around the average value or level than the male. This difference in spread is clearly indicated by the fact that the male standard deviation is more than twice as great as the female, i.e. 14.67 compared with 6.02.

Let us now take another break and stop recording our analysis with the command MTB > NOOUTFILE. A copy of our analysis should now be safely in a file called SUICIDE3.LIS on the floppy disc in A:.

Let us also save the data in the editor with the command MTB > SAVE 'A:\SUICIDE3' which will give us a file called SUICIDE3.MTW on the floppy disc in A:.

This is an appropriate place to remind you that you should study your computer print-outs carefully. Check for errors and make handwritten annotations to help you make sense of the output at a later date. Also, if fair copies of the analysis and commentary are needed, the handwritten comments form a necessary step in producing an edited final version with a word-processor.

Postscript

We have in this chapter applied various simple exploratory data analysis techniques to the data on suicide rates, first by hand and then using the Minitab package. We now know a lot more about the suicide rates and how they vary in these 16 countries. In our analysis we have focused primarily on the male data and then we made a simple comparison of male and female suicide rates by looking at the overall rates. We have not examined the female data in any depth and neither have we compared the male and female rates for the various age groups.

It would be interesting to make a comparison of the differences in the male and female rates for each country over time. However, we have probably done enough work on this data set for you to get a good idea of the simpler tools of analysis to be found in Minitab. Whilst realizing that there is much else we could do, let us suppose we wish to end our exploratory data analysis of suicide rates at this point. What remains to be done is to print out the various analysis files, such as SUICIDE2.LIS etc., study them carefully and gather together all our notes and comments into a brief report on the male suicide rates for the selected countries. Such a report, compiled from our previous work, comments and observations should of course be written in complete English sentences and not in the form of cryptic notes. If possible it should be typed. It should also give an overview of the methods of analysis used and comment on their usefulness. Some brief comment on the adequacy or reliability of the data set should also be included in the report.

Exercises

1 In this exercise I would like you to take our analysis of the men's suicide data one stage further. Earlier we looked at the process of standardization of data and worked, by hand, on the two extreme age groups. I would now like you to standardize all the age groups using Minitab. You should proceed in stages, first removing the level and then the spread.

We know from our previous analysis how the level (median and mean) of the suicide rates changes over the age cohorts. First we remove or set aside these levels (using the median only) so that other features of the distributions can be studied. After this, in the final process of standardization we remove the spread (as represented by the midspread).

You will remember that the standardized score was defined as

$$\text{standardized score} = \frac{\text{value} - \text{level}}{\text{spread}}$$

When using the median and midspread (d_q) as measures of level and midspread respectively this becomes,

$$\text{standardized score} = \frac{\text{value} - \text{median}}{\text{midspread}}$$

(i) Retrieve one of the saved suicide data files containing the male suicide rates with the command say, MTB > RETRIEVE 'A:\SUICIDE2.MTW'. Give an OUTFILE command such as MTB > OUTFILE 'A:\SUICIDE5'. You may wish to set the width of the output file with the command MTB > OW = 70; such a command would prevent a tiresome wraparound of output such as graphs and plots if we wish to print the file with a word-processor whose effective width may be 70 (i.e. a screen of 80 with margins set at 5 and 5).

Before starting on our analysis it is worth checking which columns of our file contain data with the command MTB > INFO and then clearing the Editor of columns that we do not require with the ERASE command, for instance MTB > ERASE C20-C30.

For this exercise we need the median and the upper and lower quartiles, Q3 and Q1, so we first get a statistical description of the male suicide data with the command MTB > DESCRIBE C13-C19.

As the first stage of standardization, i.e. removing the level, you should now subtract the median of each column from every value of the variable in that column with commands such as

```
MTB > LET C23=C13-16.80
MTB > LET C24=C14-24.90
MTB > LET C25=C15-28.10 etc.
```

Columns C23–C29 should now contain the age cohort distributions with the level (median) removed. Now plot all these columns, C23–C29, on one axis using the stacking technique described earlier.

```
MTB > STACK C23-C29 and put into C30
MTB > SET C31
DATA> 16(3) 16(4) 16(5) 16(6) 16(7) 16(8) 16(9)
DATA> END
```

You should now check that C30 and C31 contain an equal number of rows (16 x 7 = 112) with the command MTB > INFO.

Now plot the age cohort data with the levels (medians) removed with the commands

```
MTB > BOXPLOT C30;
SUBC> BY C31.
```

At this stage it is probably worth checking your results with the print-out given in the answers to this exercise.

In the boxplots the medians, as indicated by the cross in each box, now all appear at zero and we can clearly see what we noted previously regarding the spread of the age cohort data. The spread increases with age, although there is some slackening of this increase in the 55–64 year cohort. The oldest group of men, those of 75+ years, has the largest spread indicated by the box, whiskers and outlier.

(ii) I would now like you to go one stage further and remove the spread and so find and plot the standardized data. You should standardize using the median and midspread (d_q) and so the appropriate formula is

$$\text{standardized score} = \frac{X - \text{median}}{\text{midspread}} = \frac{X - \text{median}}{\text{Q3} - \text{Q1}}$$

We have already found the columns (C23–C29) which correspond to the top line of this equation. Now calculate the various values that go on the bottom line of the formula. Do this by subtracting the lower quartile Q1 from the upper quartile Q3 for each age cohort. For example, for C13 we have Q3 = 23.65 and Q1 = 8.4 and so the midspread Q3 – Q1 = d_q = 23.65 – 8.4 = 15.25. For C14 we have Q3 = 33.65 and Q1 = 14.75 and so the midspread d_q= Q3 – Q1= 33.65 – 14.75 = 18.9, and so on.

You should now calculate the standardized columns by dividing each of the C23–C29 by the appropriate midspread as follows:

```
MTB > LET C33=C23/15.25
MTB > LET C34=C24/18.90
```

and so on.

To get the final plot of the standardized columns C33–C39 you should stack the data and put in a control column as previously with the commands

```
MTB > STACK C33-C39 and put into C40
MTB > SET C41
DATA> 16(3) 16(4) 16(5) 16(6) 16(7) 16(8) 16(9)
DATA> END
```

You should check that you have equal numbers (16 x 7 = 112) in the control column C41 and in the column of standardized data C40 with the command

```
MTB > INFO
```

Finally you should plot the standardized data as a boxplot with the command

```
MTB > BOXPLOT C40;
SUBC> BY C41.
```

Save your data sheet with the command MTB > SAVE 'A:\SUICIDE4' and then end outfiling your analysis with the command MTB > NOOUTFILE and exit from Minitab with MTB > STOP.

(iii) Has the process of standardization, by the removal (setting aside) first of the level and then of the spread, allowed us to see any new features of the distribution of men's suicide rates?

Answers to Exercises

1 (i and ii)

```
MTB > OW=70
MTB > RETRIEVE 'A:\SUICIDE2.MTW'

Worksheet retrieved from file: A:\SUICIDE2.MTW
MTB > INFO

COLUMN   NAME       COUNT
C1                     32
C2                     32
C3                     32
C4                     32
C5                     32
C6                     32
C7                     32
C8                     32
C9                     32
C12      MALEALL       16
C13      M15-24        16
C14      M25-34        16
```

```
C15     M35-44        16
C16     M45-54        16
C17     M55-64        16
C18     M65-74        16
C19     M75+          16
C30                  128
C31                  128

CONSTANTS USED: NONE

MTB > ERASE C30 C31
MTB > DESCRIBE C13-C19

              N      MEAN    MEDIAN    TRMEAN     STDEV   SEMEAN
M15-24       16     16.34     16.80     16.21      8.12     2.03
M25-34       16     26.28     24.90     24.68     15.29     3.82
M35-44       16     28.77     28.10     25.84     19.92     4.98
M45-54       16     35.24     31.65     31.31     24.42     6.11
M55-64       16     35.56     31.65     32.47     19.16     4.79
M65-74       16     41.87     34.60     38.26     25.44     6.36
M75+         16      65.3     49.0       60.6      46.2     11.6

            MIN       MAX        Q1        Q3
M15-24     5.20     29.30      8.40     23.65
M25-34     7.70     67.30     14.75     33.65
M35-44     9.30     89.30     13.48     37.02
M45-54    13.90    111.50     15.75     45.25
M55-64    17.80     96.50     22.15     42.60
M65-74    15.80    118.40     25.73     53.50
M75+        8.8     188.0      36.8      84.8

MTB > LET C23=C13-16.80
MTB > LET C24=C14-24.90
MTB > LET C25=C15-28.10
MTB > LET C26=C16-31.65

MTB > LET C27=C17-31.65
MTB > LET C28=C18-34.60
MTB > LET C29=C19-49.00
MTB >
MTB > STACK C23-C29 put into C30
MTB > SET C31
DATA> 16(3) 16(4) 16(5) 16(6) 16(7) 16(8) 16(9)
DATA> END
MTB > INFO
```

```
COLUMN   NAME      COUNT
C1                    32
C2                    32
C3                    32
C4                    32
C5                    32
C6                    32
C7                    32
C8                    32
C9                    32
C12      MALEALL      16
C13      M15-24       16
C14      M25-34       16
C15      M35-44       16
C16      M45-54       16
C17      M55-64       16
C18      M65-74       16
C19      M75+         16
C23                   16
C24                   16
C25                   16
C26                   16
C27                   16
C28                   16
C29                   16
C30                  112
C31                  112

CONSTANTS USED: NONE

MTB > BOXPLOT C30;
SUBC> BY C31.

C31
                      -----
3                   -I + I-
                      -----
                      -----
4                  --I + I---      *
                      -----
                    -------
5                 -I   + I--        *
                    -------
                    --------
6                   I   + I---          *
                    --------
                      ------
7                   -I +  I-        *
                      ------
                      -------
8                 ---I  + I----         *
                      -------
                     ------------
9           -------I     +    I----------           *
                     ------------
        ----+---------+---------+---------+---------+------C30
         -40          0        40        80       120
```

```
MTB > #Midspreads are as follows,
MTB > #ie. Q3-Q1 for each column,
MTB > #C13 15.25, C14 18.90,
MTB > #C15 23.54, C16 29.50, C17 20.45,
MTB > #C18 27.77, C19 48.0.
MTB >
MTB > LET C33=C23/15.25
MTB > LET C34=C24/18.90
MTB > LET C35=C25/23.54
MTB > LET C36=C26/29.50
MTB > LET C37=C27/20.45
MTB > LET C38=C28/27.77
MTB > LET C39=C29/48.00
MTB >
MTB > STACK C33-C39 and put into C40
MTB > SET C41
DATA> 16(3) 16(4) 16(5) 16(6) 16(7) 16(8) 16(9)
DATA> END
MTB > INFO

COLUMN    NAME        COUNT
C1                       32
C2                       32
C3                       32
C4                       32
C5                       32
C6                       32
C7                       32
C8                       32
C9                       32
C12        MALEALL       16
C13        M15-24        16
C14        M25-34        16
C15        M35-44        16
C16        M45-54        16
C17        M55-64        16
C18        M65-74        16
C19        M75+          16
C23                      16
C24                      16
C25                      16
C26                      16
C27                      16
C28                      16
C29                      16
C30                     112
```

```
C31                  112
C33                   16
C34                   16
C35                   16
C36                   16
C37                   16
C38                   16
C39                   16
C40                  112
C41                  112

CONSTANTS USED: NONE

MTB > BOXPLOT C40;
SUBC> BY C41.
```

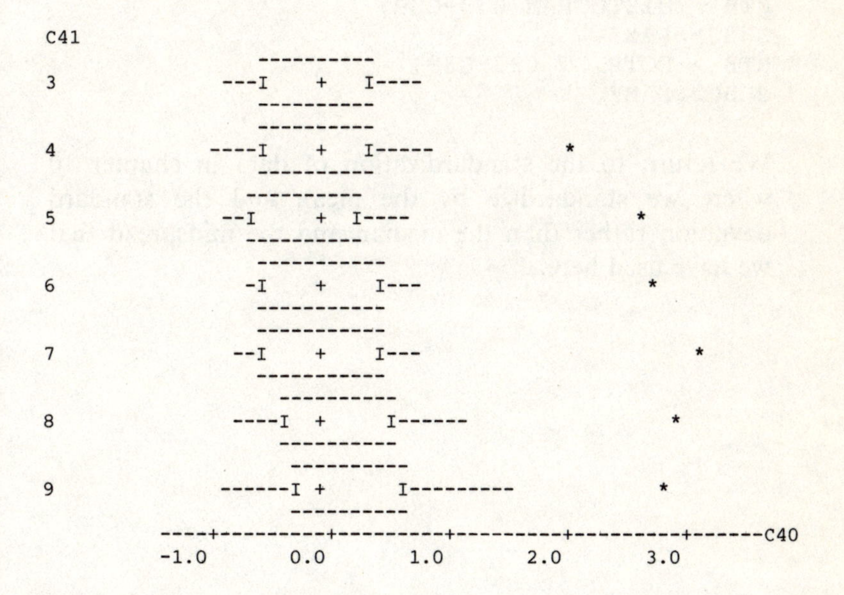


```
MTB > SAVE 'A:\SUICIDE4'
```

(iii) The first part of the process of standardization, that of the removal of the level, in our case the median, does enable us to confirm what we noted earlier. The spread of the age cohorts increases with age but there is some slackening of this increase in the cohort of the men aged 55–64 years.

However, the fully standardized plot, where both the level and the spread of the distributions are removed, does not seem to add much to our previous knowledge of the data. The removal of the level and spread of the various distributions should have enabled us to see the skewness or lopsidedness of the data more clearly. However, the rather crude representation of the data by means of boxplots does not allow us to see the skew at all clearly. In fact Minitab does not have a simple command that allows us to measure or display the skewness of data.

We can get some idea of the skewness of our data, however, by plotting the histograms or dotplots of the standardized data in columns C33–C39 using the commands

```
MTB > HISTOGRAM C33-C39;
SUBC> SAME.
MTB > DOTPLOT C33-C39;
SUBC> SAME.
```

We return to the standardization of data in chapter 10 where we standardize by the mean and the standard deviation rather than the median and the midspread that we have used here.

6 Two Variables, Correlation, Cause and Effect

In this chapter we look at some ideas on the relationship between two variables. We first consider the related variables of height and weight by looking at relatively simple data for 60 seamen and then briefly examine the notion of causality before outlining a simple but useful concept of correlation.

Sometimes in data analysis we are concerned with one variable and how it is distributed, but more often we have more than one variable and we are interested not only in how each is distributed but whether the variables are related to each other. A further question immediately follows: if the variables are connected or related to each other, how are they related? Does variable *A* always appear with variables *D* and *E*? Are *A* and *B* causally related? If so does *A* cause *B*, or *B* cause *A*? Or maybe both *A* and *B* are related to or dependent on variables *F* or *G* say, which we may not have measured, and indeed may not even be aware of.

Relations or connections between variables are often complex, especially when we attempt to decide what causes what. Often we stop short of assigning causal relationships and prefer to work out correlations between variables that often appear together. Such correlations are frequently suggestive and can help us formulate other possible relations between variables. For instance, let us suppose that we have data on perinatal mortality (stillbirths and deaths in the first week of life) and corresponding data on such variables as mother's age, social class, height, smoking habits in pregnancy and area of domicile. Is there a relationship between these various factors and the main variable of mortality? If such connections do exist how strong is each relationship? How may the strengths of such relationships, if they exist, be quantified?

Such considerations involving several variables are extremely complex and demand not only an extensive knowledge of statistics but a great deal of knowledge of the substantive area being studied. In this chapter we simplify the task by looking initially at the relation between just two variables. We examine variables that we know are linked together, i.e. the height and weight of people. We attempt to picture their correlation through scattergraphs and then look at a simple numerical measure which describes such a correlation. Later in the chapter we return to questions of relationships such as causality between variables.

Heights and Weights of Young Seamen

Our knowledge of the physical world tells us that, in general, increased height goes with or is related to increased weight. Furthermore we know that weight is calculated from volume and density and the dimension of height is directly related to volume. Let us now look at the data for the heights and weights of 60 young seamen table 6.1. These data are to be found in the file DATA05.DAT.

A distribution of such paired values of height and weight is called a bivariate distribution. Is there, in this bivariate distribution, a general relationship between weight and height? We have 60 pairs of variables and as a first step we can plot them as ordered pairs i.e. (85, 188), (69, 173), etc. on a scatter diagram. They are called ordered pairs because here order does matter. We give the pair of values in the order, weight and then height.

Table 6.1 Heights and weights of 60 seamen measured to the nearest centimetre and kilogram

Name	*kg*	*cm*	*Name*	*kg*	*cm*
Baker	85	188	Joshua	74	185
Beard	69	173	McDonald	65	178
Breedon	75	183	Maidment	69	170
Brookes	58	181	Maund	60	173
Burman	66	178	Mawford	50	175
Cameron	48	165	Middleton	61	173
Cheeseman	73	173	Mitchell	80	188
Coleman	75	191	Morgan	59	179
Coombe	62	177	Munro	67	175
Day	73	185	Newton	64	176
Doughty	46	163	Parr	67	179
Dunkley	73	178	Parry	62	177
Eldridge	58	175	1xgram	80	180
Elworthy	73	181	Percy	71	185
Ewen	75	188	Phipps	51	170
Fox	71	175	Pizzey	55	175
Gaukroger	64	183	Rentell	62	183
Gibbons	67	173	Ross	51	169
Goodsir	59	178	Ruegg	48	173
Griffiths	55	172	Sargent	68	180
Hardy	61	180	Smith	63	176
Harper	70	173	Stanness	57	170
Hart	54	179	Stares	71	173
Hawkins	68	178	Strafford	60	170
Hicks	75	187	Teagles	53	169
Hill	57	168	Tearle	68	180
Holloway	77	178	Wells	54	171
Horsfield	68	175	Woodgate	67	179
Isherwood	53	168	Youngman	71	182
Jarvis	62	178	Zambelli	63	176

We could plot the scatter diagram of the bivariate data by hand or rather more easily by reading the data file into Minitab with, MTB > READ 'A:\DATA05.DAT' into C1 C2 and then plotting the ordered pairs as follows.

Go into the editor and enter the labels Weight and Height above C1 and C2 respectively. With the command MTB > PLOT C1 against C2 you will get a scattergraph of the bivariate distribution (figure 6.1).

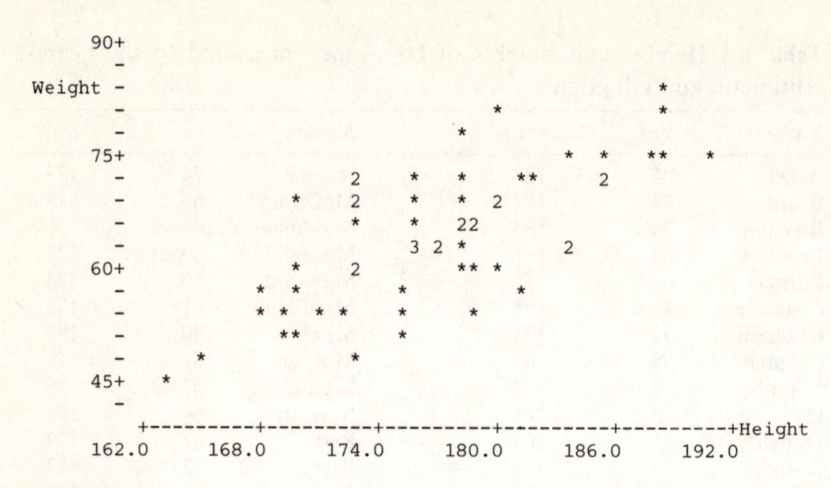

Figure 6.1

As can be seen the plotted bivariate points form a pattern of asterisks. The 2s and 3s in the scatter indicate coincident points, i.e. two or three seamen with the same height and weight. The points are scattered in a roughly elliptical or lenticular shape from lower left to upper right of the scattergraph. This indicates the general tendency in the data for low weight to be associated with low height and greater weight with greater height. That is, in general, when one variable increases, so does the other. This does not surprise us as we generally expect tall people to be heavier than short ones. However, this is not invariably the case and we have in the data a few tall yet light seamen (e.g. Brookes 58, 183) and also a few short and heavy ones (e.g. Maidment 69, 170). These few exceptions or outliers, however, serve to highlight the general rule.

This scatter diagram suggests that there is a general association between the two variables. There are two ways that this association or correlation can be expressed, first by finding the equation of a straight line that may be fitted in the general direction indicated by the pattern of points (a regression line) or second by calculating a correlation coefficient. In this chapter we look only at the way to express this type of association in the form of a rank correlation coefficient. Fitting lines to data will be dealt with in the next chapter.

Spearman's Rank Correlation Coefficient

Different types of correlation coefficient exist, the two most common being Spearman's rank correlation coefficient and Pearson's product moment correlation coefficient. In this chapter we deal only with the former which is relatively easy to calculate and will serve well enough to illustrate the concepts of correlation. It can also be used with data which are not measured on a nominal scale (such as height, weight etc.) but are merely ordered (ordinal data) in some way, for instance by personal preference or taste. Pearson's product moment coefficient is rather more difficult to calculate by hand. However, Minitab can calculate both types with equal facility. We should notice, though, that if we require the Spearman correlation coefficient in Minitab we first need to rank the data. Before looking in more detail at Spearman's rank correlation coefficient, let us consider a few scatter diagrams that might arise from different kinds of bivariate data.

Let us suppose we have a number of measurements of two variables called, say, X and Y. If we suspect that an association exists between them the first step in analysis is to draw a scatter diagram for the data, just as we did for the seamen's data. We might get one of the types of scatter diagram shown in figure 6.2. Let us briefly look at each diagram in turn.

(a) Here, as in our earlier example of heights and weights, there appears to be a roughly linear pattern to the scatter. It looks as if a straight line could probably be fitted to the data. It would have what is called a positive slope, running from lower left to upper right. It is likely that a positive linear correlation exists which would indicate that in general X increases as Y increases.

(b) Here also the scatter seems to indicate linear correlation, but in this case it is negative linear correlation. The fitted line would have what is called a negative slope and would run from top left to lower right of the scattergraph. That is, low values of one variable are associated with high values of the other. An example of this type of relation might be found, say, between human running speed and age over 40 years (but less than 50 years). Speed decreases as age increases, and if the time span (in this case 10 years) is not too long the relationship is approximately linear. This

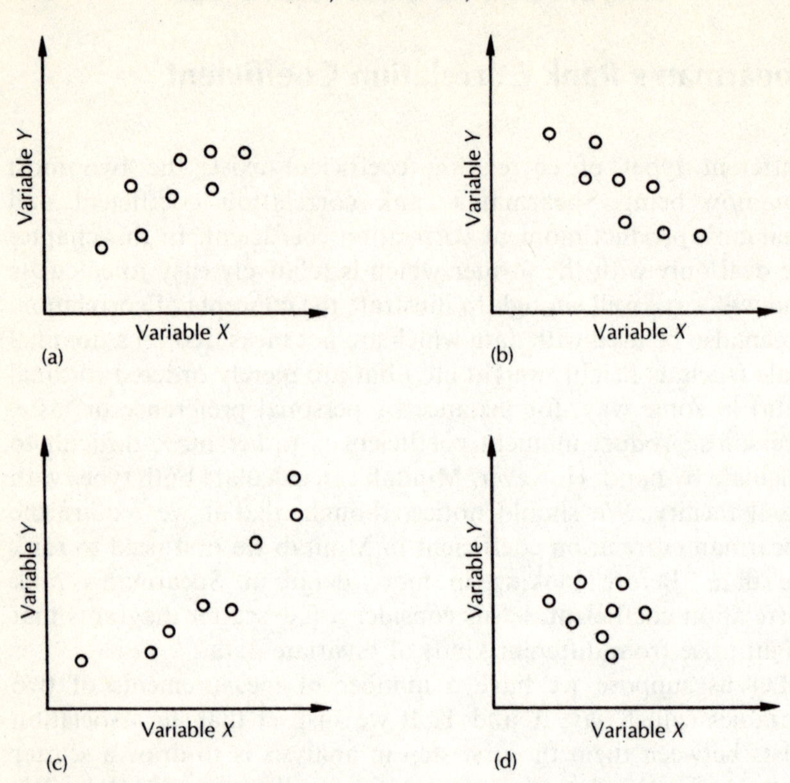

Figure 6.2

type of correlation is called a negative or inverse correlation. The correlation coefficient would be a negative number.

(c) In this scattergraph there does seem to be a pattern to the scatter of points; however, it appears to be a curvilinear rather than a linear pattern. This type of association between variables is not one we can deal with directly in terms of the linear correlation coefficient considered in this chapter. However, in a later chapter we investigate how various types of transformation will allow us to straighten out many such curves.

(d) In this case there appears to be no obvious pattern or correlation between the variables. The correlation coefficient would probably be very small and could be positive or negative. Again, as we shall see later, a transformation of the data may, in some cases, help us to detect a pattern which is not obvious at first sight.

Let us now turn from scattergraphs to the calculation of a correlation coefficient, using Spearman's rank correlation coefficient. This is one of the most frequently used measures of association and it can be used not only for nominal data where the measurement of variables is possible (e.g. as in the heights and weights data), but also for ordinal data where the data are merely ordered because precise measurement of the variables is not possible or difficult. That is, we may use this technique not only to compare bivariate data in terms of, say, height and weight which can be measured quite precisely but also to make sense of, say, a collection of artistic preferences. We could for instance compare what two groups of people thought about the artistic merit of ten paintings, or the taste of various types of beer or ice cream. It is only necessary to rank the data in a sequence and we can then compare the rankings for each group and obtain a correlation coefficient.

The Spearman rank correlation coefficient R_s can be calculated from the formula

$$R_s = 1 - \frac{6\sum d^2}{n(n^2 - 1)}$$

where n is the number of pairs ranked and d is the difference between corresponding ranks. The sigma sign $\sum$ indicates that we must sum the squared differences in rank.

The calculated value of R_s will vary between $+1$ and -1, which indicate perfect positive and negative correlation respectively, between the sets of figures. The nearer the value of R_s is to either $+1$ or -1 the stronger is likely to be the mathematical correlation between the data sets and the more likely it is that the result is significant if there is in fact a relationship between the two variables correlated.

It should be remembered that such a correlation technique can be used on any two sets of data which have an equal number of elements, but it only makes sense to correlate sets which we consider to be related to each other. To do otherwise may well result in high correlation coefficients but they will be spurious. It is well to remember that correlation coefficients indicate correlation between sets of numbers or ranks and not necessarily between any variables that these numbers or ranks represent. We return to this important point later.

Given that the correlation is not spurious, the significance of the value of R_s will depend on the number of pairs used in its calculation. For example, an R_s calculated to be $+0.8$ for 10 pairs, would not represent as high a degree of mathematical correlation as say an R_s of $+0.8$ for 50 pairs.

The 60 pairs of seamen's heights and weights that we used at the beginning of this chapter produce a Spearman's rank correlation coefficient R_s of approximately $+0.7$. This is a high correlation and is likely to be significant at the 0.99 or 99% level of probability. That is, given that there is a relationship between the variables of height and weight, such a correlation is unlikely to occur by chance in 99% of such cases collected at random.

Once the Spearman coefficient has been calculated it must always be compared with a table of critical values (Appendix 2, Table 5) at, say, the 95% or 99% level for various values of n (n is the number of pairs correlated). The R_s correlation coefficient (or test statistic) will be accepted if it is equal to or greater than the critical value and rejected if it is less. In making this comparison between the calculated R_s and the critical values in Neave, the sign (i.e. + or –) of the coefficient is ignored. You should use the critical values in the row marked α_2. Let us now consider a simple example.

Example 1 The percentage examination scores of 10 students in data analysis and economics were as follows. Calculate the Spearman rank correlation coefficient for these data.

Student	*A*	*B*	*C*	*D*	*E*	*F*	*G*	*H*	*I*	*J*
Data analysis	65	90	52	44	95	36	48	63	80	15
Economics	62	71	58	58	64	40	42	66	67	55

A scatter diagram will give a preliminary indication of whether linear correlation exists, and so we plot the ordered pairs (65, 62), (90, 71) etc. The amount of data is fairly small so we can probably plot the graph by hand (figure 6.3).

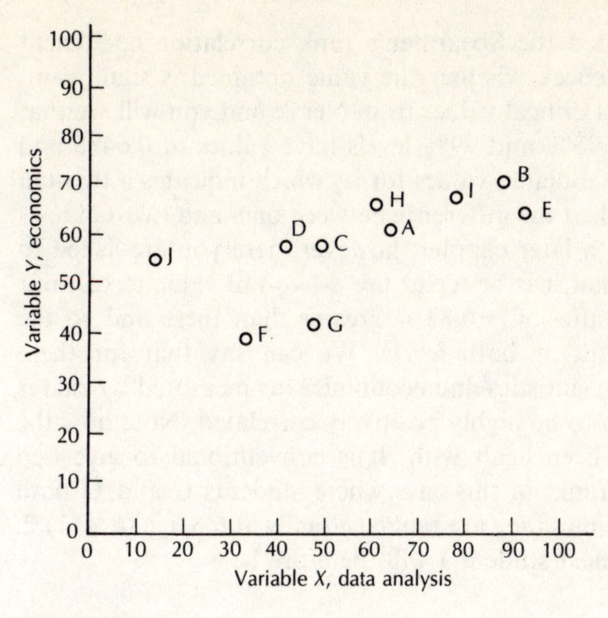

Figure 6.3

It appears likely from the scatter diagram that some form of positive correlation exists. We will now rank the two sets of data (table 6.2) and calculate R_s.

Table 6.2 Examination scores in two subjects

Student	*Data analysis (X)*	*Economics (Y)*	*Rank X*	*Rank Y*	*d*	*d^2*
A	65	62	4	5	1	1
B	90	71	2	1	1	1
X	52	58	6	6.5	0.5	0.25
D	44	58	8	6.5	1.5	2.25
E	95	64	1	4	3	9
F	36	40	9	10	1	1
G	48	42	7	9	2	4
H	63	66	5	3	2	4
I	80	67	3	2	1	1
J	15	55	10	8	2	4
						$\sum d^2 = 27.5$

$$R_s = 1 - \frac{6\sum d^2}{n(n^2-1)} = 1 - \frac{6 \times 27.5}{10(100-1)} = 1 - 0.167 = +0.83$$

Having calculated the Spearman's rank correlation coefficient we now need to check whether the value obtained is significant. Check the table of critical values from Neave and you will see that with $n = 10$ the 95% and 99% levels have values of 0.6485 and 0.7939. (Use the tabulated values for α_2 which indicates a two-tail test. We shall look at the difference between one- and two-tail tests of significance in a later chapter; however, here you are asked to take it on trust that it is better to use a two-tail value to test our result.) Our R_s value of $+0.83$ is greater than these and so the result is significant at both levels. We can say that for these students ability in statistics and economics (as measured by exams anyway!) appears to be highly positively correlated. Note how the tied scores have been dealt with. It is conventional to give tied scores the same rank; in this case where students C and D both have 58 in Economics they are ranked equally at 6.5, i.e. $(6 + 7)/2$. The rank of the next student J will therefore be 8.

Interpretation of the Correlation Coefficient

We need to be aware that a high value of the coefficient R_s does not prove beyond doubt that a strong relationship exists (except in a mathematical sense between two sets of numbers). If we are using say a 0.95 critical level and our calculated value of R_s only just exceeds it, there is still a chance in five cases out of 100 that such conjunctions could arise by chance. If we need a greater degree of certainty we must increase n, the number of pairs compared.

However, there is a much more important consideration that we need to stress. Even where a high (positive or negative) correlation exists which is well above the chosen critical level of 95% or 99% this may indicate a high mathematical correlation between the ranks but does not necessarily mean that the *variables themselves* are correlated or even connected, and still less that one variable causes the other. The technique we have looked at measures the mathematical relationship between the two sets of figures and measures the significance of this by comparing it with what might have arisen randomly by chance. The technique tells us *nothing* about the kind of relationship that may or may not exist between the variables involved. Such relationships and imputed causal relationships must be established quite separately from such computation. Knowledge of such relationships must come not from the world of mathematical concepts but from the 'real' social or natural world. That is, we need some independent reason outside mere lists of figures to establish relationships between variables. A considerable knowledge of the field from which the statistics have been drawn is usually necessary. The calculation of correlation coefficients is extremely useful and may help us confirm or disprove beliefs or hunches, but they do not and cannot in themselves constitute firm evidence of relationship.

High correlations may be entirely spurious and the variables compared quite unconnected. Or it may be that highly correlated sequences of numbers may only be tenuously connected by some third factor. For instance if we find that in a certain British city the sales of video-recorders and canned beer is highly positively correlated, there may be some connection between the two variables in that watching films at home may involve entertaining friends with canned beer. However, it is likely that the sales of *both* types of consumer good reflect more basic economic or social factors.

The point to remember is that high (positive or negative) correlations do not necessarily indicate a relationship and still less a causal relationship between the variables considered. Meaningful relations and causation are complex ideas and evidence and judgement outside analytical statistical considerations are necessary. The examples in the exercise at the end of the chapter illustrate some of these complexities.

We should also note that a value of R_s which is near zero and not significant does not necessarily indicate that the two variables are unconnected, but merely that there is no *linear* relationship between them. There may well be a curvilinear or other complex relationship between the variables, however, which Spearman's rather blunt instrument will not detect. Hence it is always useful to draw a scatter diagram of the data and draw preliminary conclusions from that. Let us now look at another example.

Example 2 Look at the following data on divorce and birth in England and Wales (table 6.3). Plot the data on a scatter diagram using Minitab and then calculate Spearman's rank correlation coefficient (i) by hand and (ii) by using Minitab.

Table 6.3 Divorce and birth rates

Year	*Divorce (X)*	*Births (Y)*
1968	3.7	819
1969	4.1	798
1970	4.7	784
1971	6.0	783
1972	9.5	725
1973	8.4	676
1974	9.0	640
1975	9.6	603
1976	10.1	584
1977	10.4	569

The figure for divorce is the rate per 1000 in the married population. The births are the total births in thousands

Source: CSO, 1980

(i) Let us calculate the Spearman's rank correlation coefficient by hand as follows (table 6.4):

Table 6.4 Rankings of divorce and births

Year	*Rank X*	*Rank Y*	*d*	d^2
1968	1	10	9	81
1969	2	9	7	49
1970	3	8	5	25
1971	4	7	3	9
1972	7	6	1	1
1973	5	5	0	0
1974	6	4	2	4
1975	8	3	5	25
1976	9	2	7	49
1977	10	1	9	81
				$\sum d^2 = 324$

$$R_s = 1 - \frac{6\sum d^2}{n(n^2 - 1)} = 1 - \frac{6 \times 324}{10(100 - 1)} = 1 - 1.96 = -0.96$$

(ii) After entering the figures for divorce and birth into columns C1 and C2 respectively in the Editor and then naming the columns we can use the command MTB > PLOT C2 against C1. To calculate the Spearman correlation coefficient using Minitab we first need to rank the data on birth and divorce with the commands,

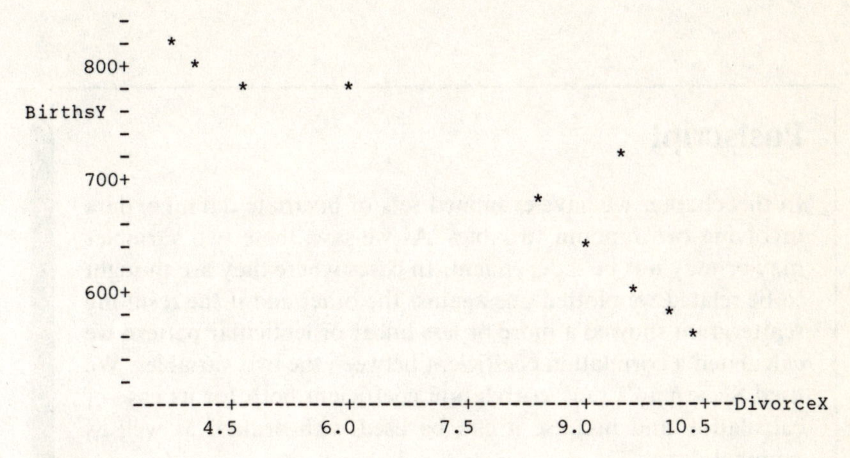

Figure 6.4

```
MTB > RANK C1 and put into C3
MTB > RANK C2 and put into C4
```

After ranking the data we can now use the command

```
MTB > CORRELATE C3 and C4
```

This gives us a Spearman rank correlation coefficient of –0.964.

It is worth noting that the Pearson product moment correlation coefficient is obtained with the command MTB > CORRELATE C1 and C2. This gives a similar correlation coefficient of –0.919 between divorce and births.

The critical values for R_s when $n = 10$ are 0.6485 (95% level) and 0.7939 (99% level). Therefore the negative correlation of –0.96 (Spearman) that we have calculated is highly significant at better than the 0.99 level. There is a very small probability that these figures are due to chance. The figures for divorce and births are strongly negatively correlated. Despite this fact and our knowledge that most children in England and Wales in this period were born in wedlock, the precise relation between the demographic factors of birth and divorce is a complex one with many intervening and confounding variables. The correlation is suggestive but the nature of the connection or relationship is one which needs to be carefully studied and explained outside such simple statistics.

Postscript

In this chapter we have examined sets of bivariate data, i.e. data involving two random variables. As we saw, these two variables may or may not be independent. In cases where they are thought to be related we plotted one against the other and if the resulting scattergram showed a more or less linear or lenticular pattern we calculated a correlation coefficient between the two variables. We used Spearman's rank correlation coefficient both for its ease of calculation and because it can be used with ordinal as well as nominal data.

It was stressed that high positive or negative correlations and correspondingly tight linear scattergrams do not necessarily indicate that a relationship exists between the variables. The variables may be unrelated and the correlations spurious. The nature of the relationship, if any, between the variables needs careful investigation. If a connection is thought to exist it may be a direct or indirect causal relationship.

In the next chapter we extend our investigation of bivariate distributions and learn how we can fit a straight line to suitably linear data.

Exercises

1 Mario and Polidori are two experienced judges of ice cream. They award the following marks out of 10 to competitors at an exhibition.

Competitors	A	B	C	D	E	F	G	H	I	J
Mario	5	2	5	8	1	7	4	9	3	10
Polidori	1	7	6	10	3	5	3	8	2	9

Plot a scattergraph either by hand or in Minitab and calculate the Spearman rank correlation coefficient. Can we say that there is significant agreement between Mario and Polidori? (Use a two-tail test.)

2 Calculate Spearman's rank correlation coefficient for the following sets of data in table 6.5:

Table 6.5 Abstract of mortality by social class and cause of death. standard mortality ratios of men, 15–64 years, 1970–1972:

Cause of death	Social class					
	1	2	3N	3M	4	5
(A) Malignant neoplasms	75	80	91	113	116	131
(B) Myocardial infarction	88	92	115	107	108	108
(C) Accident, poisoning and violence	78	78	83	94	122	197

A standardized mortality ratio of less than 100 indicates a favourable death rate for the class concerned while a value of more than 100 is unfavourable.

Source: *The Registrar Generals Decennial Supplement, England and Wales, 1978*

(a) A and B
(b) A and C

What do you think such correlations can tell us? (Use two-tail tests.)

3 Calculate Spearman's rank correlation coefficient R_s for the data in table 6.6 and briefly comment on your results. (Use a two-tail test.)

Table 6.6

Year	*1970*	*1971*	*1972*	*1973*	*1974*	*1975*	*1976*	*1977*	*1978*
U.K index of beer consumption	90	89	91	97	97	100	102	102	105
Offenders found guilty of drunkenness, England and Wales (thousands)	79	83	88	97	98	100	103	103	101

Source: *Social Trends*, 1981

4 Plot a scattergraph and a tagged scattergraph for the data in table 6.7 on nursing density and infant mortality. Also calculate R_s for the data and comment on your results.

Table 6.7

Country	*Nurses (X)*	*Infant mortality (Y)*
England and Wales	37.5	14.2
Scotland	48.2	14.8
Sweden	71.1	8.3
Norway	73.6	11.1
Denmark	80.4	10.3
Finland	81.9	11.0
Netherlands	32.2	10.6
France	50.2	11.1
West Germany	35.9	19.7
USA	63.7	16.1

X is the number of nurses in thousands per 10,000 population, 1975. Infant mortality is per 1000 live births, 1975.

Source: DHSS, 1980

5 In this exercise I would like you to look at some data abstracted from a UNICEF publication, *The State of the World's Children*. These data are contained in DATA11.DAT and you should look at the file in appendix 1.

(i) Read the data into the Minitab data Editor with the command MTB > READ 'A:\DATA11.DAT' into C1-C12. Open an OUTFILE with a command such as MTB > OUTFILE 'A:\CHILDRN1'

Let us focus our attention on three variables in the data file, and label them in the data Editor as follows:

Annual gross national product per capita (C2)	GNP
Infant mortality rate (C7)	IMR
Maternal mortality rate (C10)	MMR

We will also use the control column (C1) which categorizes the countries by the mortality rate of children under 5 years of age (U5MR).

(ii) Obtain a description and boxplot of the GNP per head with the commands

```
MTB > DESCRIBE C2;
SUBC> BY C1.
```

and

```
MTB > BOXPLOT C2;
SUBC> BY C1.
```

Comment briefly on each presentation of data.

(iii) Obtain a description and boxplot of the infant mortality rate with the commands

```
MTB > DESCRIBE C7;
SUBC> BY C1.
```

and

```
MTB > BOXPLOT C7;
SUBC> BY C1.
```

Comment briefly on each presentation of data.

(iv) Obtain a scattergraph of the GNP and the IMR with the command MTB > PLOT C2 against C7. Do you think that there is evidence of correlation between these two variables?

(v) Obtain a scattergraph of the IMR and the MMR with the command MTB > PLOT C7 against C10. Comment briefly on the plot. Do you think we would be justified in calculating a correlation coefficient for this bivariate distribution?

Answers to Exercises

1

```
      -                                            *
      -
   9.0+                                       *
      -
Mario -                                                   *
      -                        *
      -
   6.0+
      -
      -    *                        *
      -              *
      -
   3.0+         *
      -
      -                                  *
      -              *
      -
        --------+---------+---------+---------+---------+--Polidori
              2.0       4.0       6.0       8.0      10.0
```

	A	*B*	*C*	*D*	*E*	*F*	*G*	*H*	*I*	*J*
Mario	5.5	9	5.5	3	10	4	7	2	8	1
Polidori	10	4	5	1	7.5	6	7.5	3	9	2
d	4.5	5	0.5	2	2.5	2	0.5	1	1	1
d^2	20.25	25	0.25	4	6.25	4	0.25	1	1	1

$$R_s = 1 - \frac{6 \times 63.0}{10\ (99)} = +0.62 \qquad (+0.616 \text{ using Minitab})$$

The critical value at the 95% level ($n = 10$) is 0.6485 (two-tail test). Our calculated value of R_s is less than this. There does not appear to be significant agreement between Mario and Polidori at this 95% level.

2

(a)	*1*	*2*	*3N*	*3M*	*4*	*5*
(A) Neoplasms	6	5	4	3	2	1
(B) Heart	6	5	1	4	2.5	2.5
d	0	0	3	1	0.5	1.5
d^2	0	0	9	1	0.25	2.25

$$R_s = 1 - \frac{6 \times 12.5}{6 \times 35} = +\ 0.64$$

The critical value at the 95% level (n = 6) = 0.8857. Our calculated R_s does not appear to be significant at this level.

(b)	*1*	*2*	*3N*	*3M*	*4*	*5*
(A) Neoplasms	6	5	4	3	2	1
(C) accident	5.5	5.5	4	3	2	1
d	0.5	0.5	0	0	0	0
d^2	0.25	0.25	0	0	0	0

$$R_s = 1 - \frac{6 \times 0.5}{6 \times 35} = +\ 0.986$$

Our calculated R_s of 0.986 is greater than 0.8857 and is significant at greater than the 95% level. The two sets of figures are correlated and are probably connected through class factors. However, the nature of these connections is likely to be very complex as the categories of class are very large and contain many social, educational, economic and work-related factors.

3

Year	70	71	72	73	74	75	76	77	78
Beer consumption	2	1	3	4.5	4.5	6	7.6	7.5	9
Offenders	1	2	3	4 5	6	8.5	8.5	7	
d	1	1	0	0.5	0.5	0	1	1	2
d^2	1	1	0	0.25	0.25	0	1	1	4

$$R_s = 1 - \frac{6 \times 8.5}{9(81.1)} = 2 - 0.07 = +\ 0.93$$

The critical value at the 95% level (n = 9) is 0.70, and at the 99% level (n = 10) is 0.83. Our calculated value of R_s of +0.93 is greater than both critical levels. From this general evidence the two sets of figures appear to be correlated. Higher beer consumption appears to be associated with a higher number of people found guilty of drunkenness. Probably higher beer consumption also indicates a greater amount spent on alcohol in general.

4

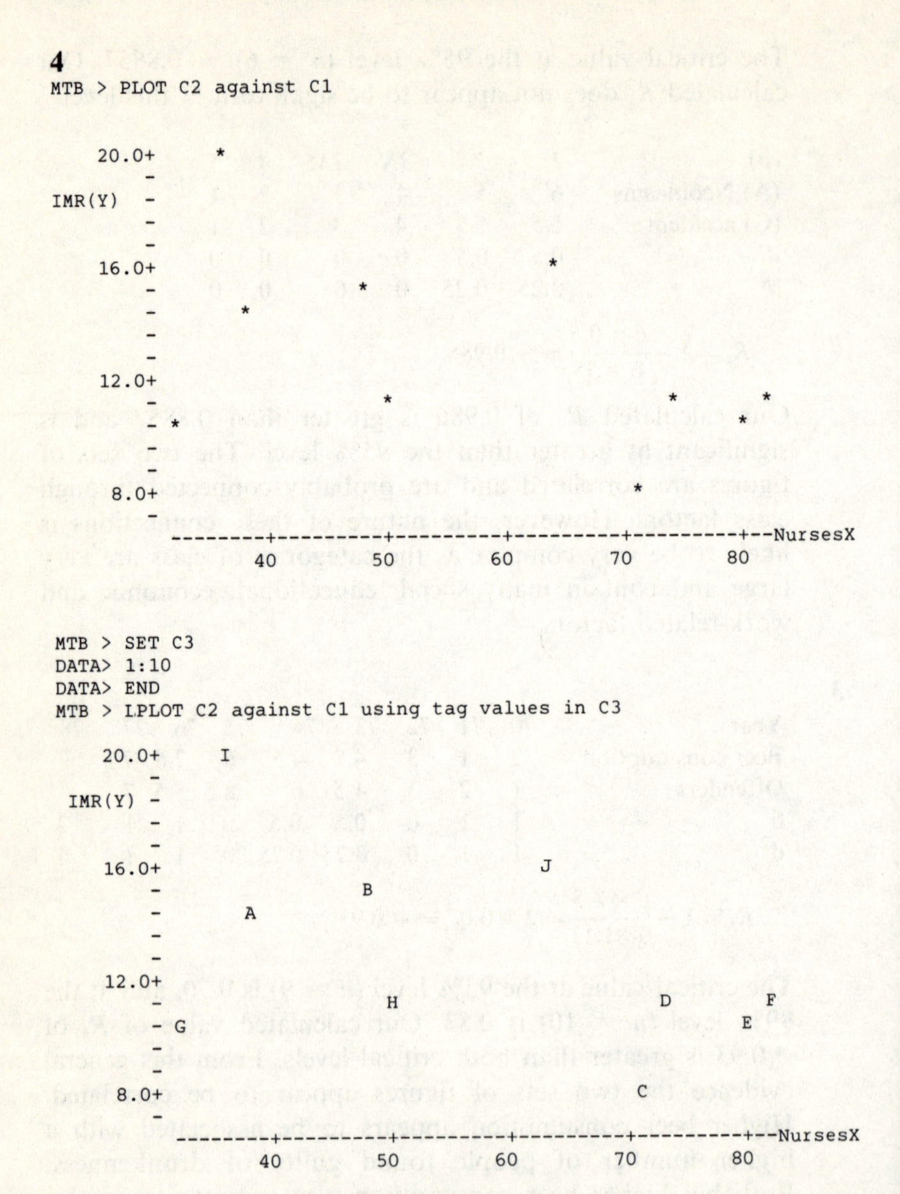

The tag values in the graph are A, England and Wales; B, Scotland; C, Sweden; D, Norway; E, Denmark; F, Finland; G, Netherlands; H, France; I, West Germany; J, USA.

	E&W	*Sc*	*Sw*	*No*	*De*	*Fi*	*Ne*	*Fr*	*Ge*	*USA*
Nurses	8	7	4	3	2	1	10	6	9	5
IMR	4	3	10	5.5	9	7	8	5.5	1	2
d	4	4	6	2.5	7	6	2	0.5	8	3
d^2	16	16	36	6.25	49	36	4	0.25	64	9

$$R_s = 1 - \frac{6 \times 236.5}{10 \times 99} = 1 - 1.43 = -0.43$$

The critical value at the 95% level ($n = 10$) is 0.6485. Therefore our figures do not indicate a significant negative correlation between the density of nurses and infant mortality. We may surmise, however, that there may be some connection between the two variables in that probably both variables are related to the amount spent on health services.

5 (i), (ii)

```
MTB > OW=70
MTB > READ 'A:\DATA11.DAT' into C1-C12
      59 ROWS READ

ROW    C1    C2    C3   C4   C5   C6    C7   C8   C9   C10

  1     1    80    5    10   35   17   173   45   21   300
  2     1   610    6    15   34    *   173   56   29     *
  3     1   180    7    11    7   17   144    *    *   170
  4     1   120    *     *    *   34   130    *    *     *
 .   .   .

ROW    C11    C12

  1     48   15.7
  2     46   10.0
  3     48    8.8
  4     46   49.2
 .   .   .

MTB > DESCRIBE C2;
SUBC> BY C1.

                C1      N     MEAN    MEDIAN    TRMEAN
GNP              1     13    286.9     250.0     275.5
                 2     15     1807      1010      1594
                 3     14     2627      1450      1509
                 4     17    14346     14610     14189
```

```
            C1   STDEV   SEMEAN        MIN       MAX
GNP          1   176.8     49.0       80.0     620.0
             2    1766      456        360      6020
             3    4623     1236        240     18430
             4    8213     1992       1170     29880


            C1      Q1       Q3
GNP          1   155.0    395.0
             2     500     2470
             3     640     2160
             4    9020    20680


MTB > BOXPLOT C2;
SUBC> BY C1.

C1
              --
1             +I
              --

              ----
2             I+ I **
              ----

              ----
3            -I+ I- O
              ----

                           -------------------
4              --------------I        +         I----------------
                           -------------------

              +---------+---------+---------+---------+---------+GNP
              0      6000     12000     18000     24000     30000


MTB > #Comment. The mean annual GNP per capita in
MTB > #dollars for the group of countries
MTB > #increases from the very low figure of
MTB > #$286.9 for group 1 countries to $14,346
MTB > #for the developed countries in group 4. We
MTB > #note that there is not only great disparity
MTB > #between the level (the mean) of the GNP
MTB > #but also the tabulated standard deviation
MTB > #and the boxplots indicate that the spread
MTB > #of GNP within each group increases as we
MTB > #go from the #less developed to the more
MTB > #developed countries.
```

(iii)

```
MTB > DESCRIBE C7;
SUBC> BY C1.

         C1        N      MEAN    MEDIAN    TRMEAN
IMR       1       13    120.23    104.00    117.82
          2       15     69.80     68.00     69.31
          3       14     34.57     35.00     34.58
          4       17     9.000     8.000     8.800

         C1    STDEV    SEMEAN       MIN       MAX
IMR       1    27.35      7.59     94.00    173.00
          2     7.75      2.00     60.00     86.00
          3    10.12      2.70     20.00     49.00
          4    2.850     0.691     5.000    16.000

         C1       Q1        Q3
IMR       1   101.50    137.00
          2    63.00     75.00
          3    25.50     43.75
          4    7.500    10.000

MTB > BOXPLOT C7;
SUBC> BY C1.

C1

                                     ---------
1                                  --I+     I---- *
                                     ---------

                            ----
2                          -I+ I----
                            ----

                   ------
3                 -I  + I--
                   ------

              --
4            -+I**
              --
            +---------+---------+---------+---------+---------+IMR
            0        35        70       105       140       175

MTB > #Here we notice a pattern in the levels and
MTB > #spreads of the infant mortality rate which is a
MTB > #kind of inverse of the pattern that we observed
MTB > #in the case of the distribution of the GNPs. The
MTB > #less developed countries of group 1 not only have
MTB > #a much higher mean rate of infant mortality but
MTB > #their spread is also much greater than that of
MTB > #the developed countries.The developed countries
MTB > #of group 4 have a very low mean rate of infant
MTB > #mortality of 9 and the #distribution for this
MTB > #group is relatively closely packed about this mean.
```

(iv)

```
MTB > PLOT C7 against C2
```

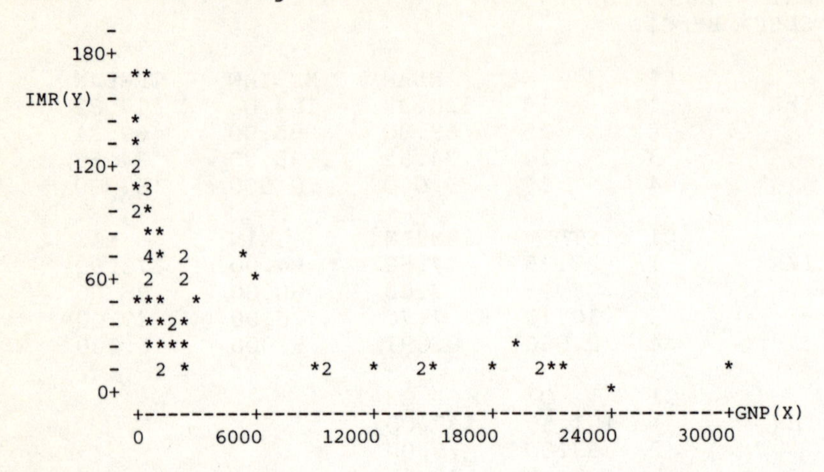


```
MTB > #There is certainly evidence of correlation
MTB > #of some kind between the two variables GNP
MTB > #and IMR, but as we can see it is not a
MTB > #linear type of relationship. The
MTB > #correlation is negative or inverse and the
MTB > #pattern is very strong, but as it is
MTB > #curvilinear we cannot use the simple
MTB > #technique of linear correlation dealt with
MTB > #in this chapter. However, as the pattern is
MTB > #very distinctive we may find that we can
MTB > #use techniques described later in chapter 8
MTB > #to 'straighten out' the data so that we can
MTB > #use linear correlation.
```

(v)

```
MTB > PLOT C7 against C10
```

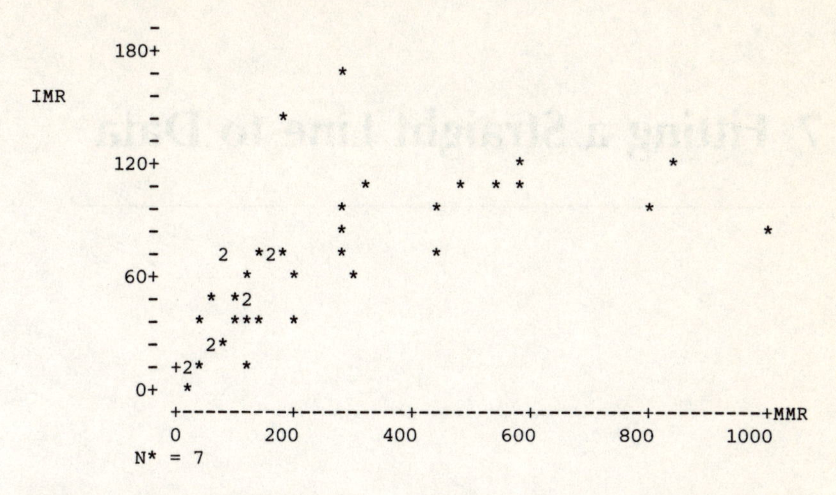


```
MTB > #We would expect there to be a pattern in
MTB > #the plot of the two variables as both are
MTB > #almost certainly related to a third
MTB > #variable, the amount spent per capita on
MTB > #health care in each country. It does seem
MTB > #from the pattern that we can see in the
MTB > #plot that there is a relationship between
MTB > #the variables MMR and IMR. The great mass
MTB > #of the data does appear to have a roughly
MTB > #lenticular or linear form and we could
MTB > #probably fit a straight line to it;
MTB > #however, there are some unusual values or
MTB > #outliers. These values would need to be
MTB > #investigated. On balance we would probably
MTB > #be justified in calculating a correlation
MTB > #coefficient here.
```

7 Fitting a Straight Line to Data

This chapter is concerned with fitting a straight-line graph to a data set. We use a method known as the least-squares regression method to find the line of optimum fit. To get a feel for the technique we fit a straight line to a small data set by 'hand' and then get Minitab to help us.

First let us consider two questions. Why do we wish to fit a straight line to data? What kind of data are suitable for such treatment?

Let us start with the last question. As we saw in the last chapter when we looked at the correlation of bivariate data, points when plotted on a scattergraph often appear to be patterned in a way which suggests that a straight line may be drawn to indicate a general trend. In some cases it might be fairly easy to fit a reasonable straight line by eye, especially if there are not many points to consider and a strong underlying linear pattern seems evident. Such a trend could also be indicated by a high positive or negative correlation coefficient between the variables. Where the data pattern seems more random in its scatter, or where an underlying curvilinear pattern is suggested, we would be unwise to try to fit a straight line to the data. In some such cases, however, a simple transformation of the data may lead to a new scatter pattern to which a straight line can be fitted. We look at such transformations in the next chapter.

The data on the heights and weights of seamen that we looked at in the last chapter have a high rank correlation coefficient ($R_s = +0.7$, $n = 60$) and when we look at the scattergraph of weight against height we may feel we could easily draw a straight line through the data to indicate its trend. Such lines drawn by eye are useful but, as you can imagine, different people might draw different straight lines. In this chapter we look at one common technique for drawing a line of best fit through a set of points. There are other techniques of line fitting but in this introductory text it seems wise to stick to just one. It is worth noting that different methods are based on different mathematical assumptions, and although the resulting fitted lines may be similar they will not, except in special circumstances, be identical.

Let us look briefly at the other question. Why do we wish to fit a straight line to data? A linear pattern is one of the most evident and straightforward patterns we can find in data and if we can find a suitable line to illustrate the pattern we will have a precise and conventional shorthand for describing the data. In this sense the line, like the measures of level and spread we looked at earlier, will help us to summarize a mass of data in a succinct way. A line may give us a brief and clear summary of a mass of data. Such a summary can help us predict values of one variable when we know the other. If, for instance, we are able to fit a straight line to the weights and heights of the seamen we could make a prediction of the height of a person in the group given his weight. Or if the sample is representative of a larger population we may be able to use the line not only to summarize the sample data but also to use it for wider predictions. Such straight-line summaries can also be used to compare the characteristics of different distributions.

Straight-line Graphs

Before dealing with the chosen fitting technique for straight lines, the least-squares regression method, we need to look at some general characteristics of straight-line graphs and the relationship between data, fits and residuals.

The data we are dealing with are bivariate, i.e. there are two variables usually labelled *X* and *Y* and we can plot the individual data points on a simple graph with a 90° degree angle between the two axes. The vertical axis running up the page is conventionally called the *Y* axis and the one running from left to right across the page is the *X* axis. The *X* variable is usually called the independent or explanatory variable and *Y* is the dependent or response variable.

The equation of a straight line is given in the form

$$Y = aX + b$$

where *a* is the gradient or slope of the line and *b* is the intercept of the line on the *Y* axis (i.e. when $x = 0$ the term aX is also zero and $y = b$).

Each bivariate point can be represented on a graph by an ordered pair or co-ordinates (x, y). As with all ordered pairs, order does of course matter and the values must be given as indicated: first the *x* value and then the *y* value. It is usual to refer to the independent and response variables using capital letters *X* and *Y* and to any individual values of the variables using lower-case letters *x* and *y* (figure 7.1). Where the need arises we will also use the suffix notation developed in chapter 4.

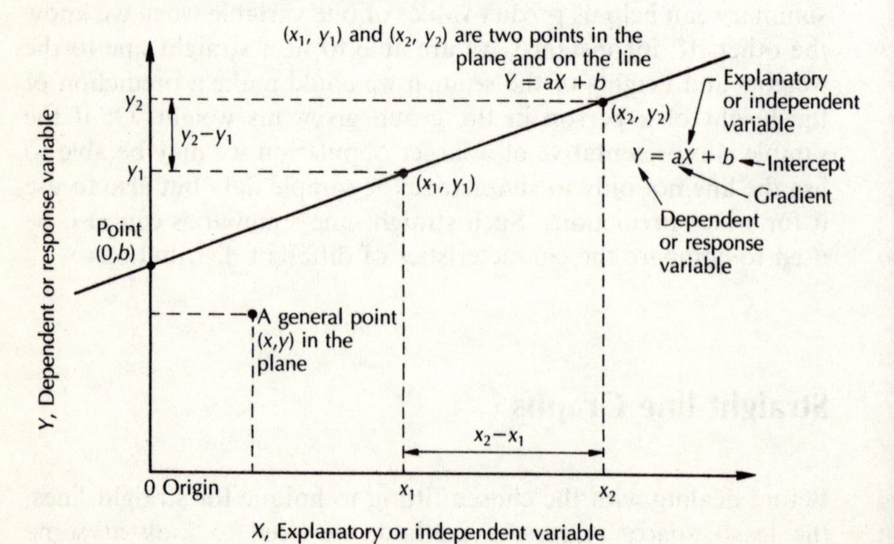

Figure 7.1

Let us now look more closely at the intercept b and the gradient a. The intercept b is found where the line cuts the Y axis. At this point the variable X is zero and the point of intercept has coordinates $(0, b)$.

The gradient a of the line $Y = aX + b$ is found from the ratio of its vertical rise to the corresponding horizontal run. It is the ratio of the change in Y to the corresponding change in X (figure 7.2).

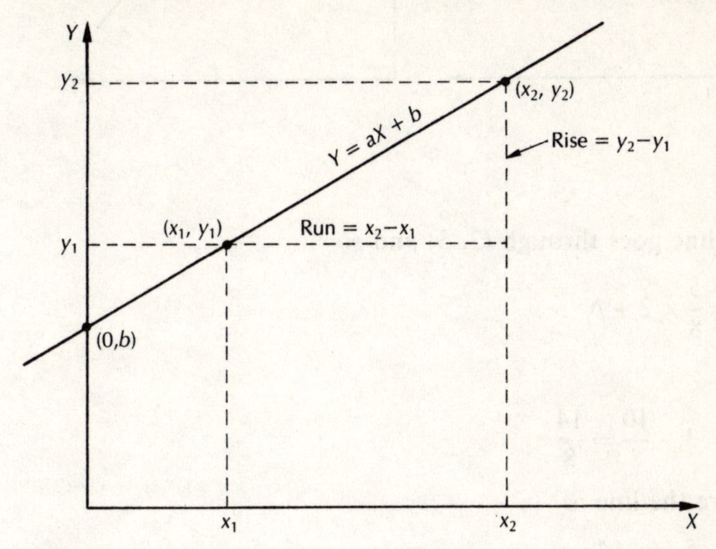

Figure 7.2

$$a = \text{gradient} = \frac{\text{vertical rise } A \text{ to } B}{\text{horizontal run } A \text{ to } B} = \frac{y_2 - y_1}{x_2 - x_1} = \frac{\text{change of } Y}{\text{change of } X}$$

What is illustrated in figure 7.2 is a positive gradient and we can of course have negative gradients which can be found from the coordinates in a similar manner. Some examples should make this clear. Consider figure 7.3(a)

$$\text{Gradient} = \frac{y_2 - y_1}{x_2 - x_1} = \frac{8 - 3}{10 - 2} = \frac{5}{8} = a$$

Therefore the line is

$$Y = \frac{5}{8}x + b$$

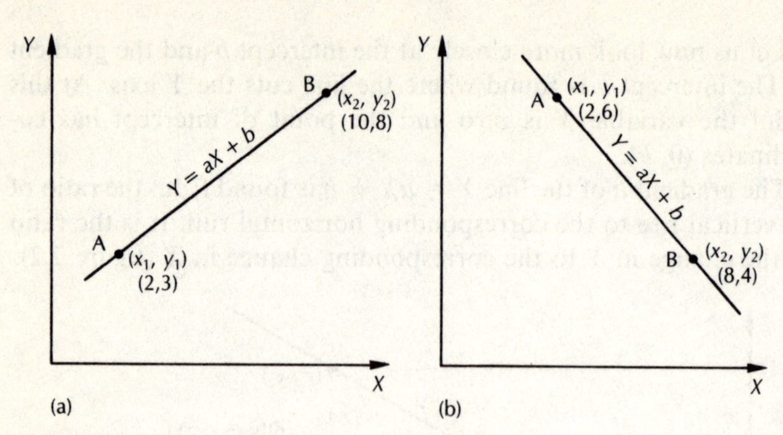

Figure 7.3

But the line goes through (2, 3) and so

$$3 = \frac{5}{8} \times 2 + b$$

and

$$b = 3 - \frac{10}{8} = \frac{14}{8}$$

Therefore the line is

$$Y = \frac{5}{8} X + \frac{14}{8}$$

Multiplying each side by 8 we have

$$8Y = 5X + 14$$

Consider figure 7.3 (b)

$$\text{Gradient} = \frac{y_2 - y_1}{x_2 - x_1} = \frac{4 - 6}{8 - 2} = \frac{-2}{6} = \frac{-1}{3} = a$$

Therefore the line is

$$Y = \frac{-1}{3} X + b$$

But the line goes through (2, 6) and so

$$6 = \frac{-1}{3} \times 2 + b$$

and

$$b = 6 + \frac{2}{3} = \frac{20}{3}$$

Therefore the line is

$$Y = \frac{-1}{3}X + \frac{20}{3}$$

Multipling each side by 3 we have

$$3Y = -X + 62$$

Data, Fit and Residuals

Let us now look in a general way at the fitting of a straight line to a set of data points. In data analysis we are dealing with empirical data and we would not expect a straight line which we wished to fit to go through each data point. The line when fitted will pass between the data points – some will be above it and some below it. We will separate the *Y* value of each data point (*x*, *y*) into a part called the fit and a part called the residual. The fit is the part that is predicted by the line we calculate and the residual is the part that remains when we compare each empirical data point with the line that has been fitted (figure 7.4). Here the fit, data and residuals are all measured vertically, i.e. in terms of the response variable *Y* at given values of *X*.

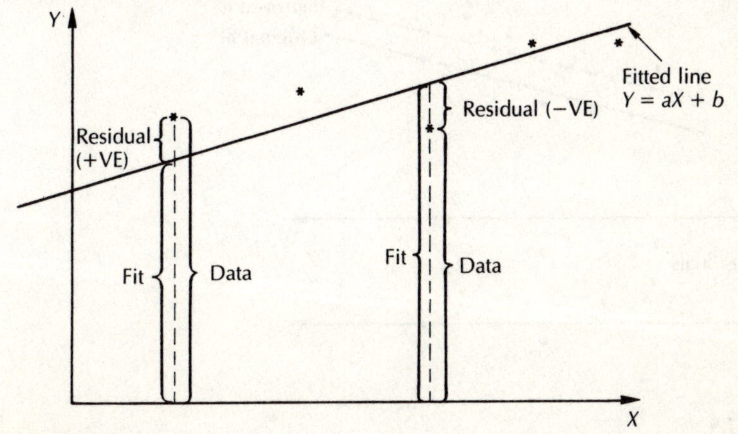

Figure 7.4

In figure 7.4 each asterisk indicates an empirical or observed data point. As can be seen,

data = fit + residual

i.e.

$$Y_{\text{data}} = Y_{\text{fit}} + \text{residual}$$

$$\text{i.e. residual} = Y_{\text{data}} - Y_{\text{fit}}$$

It is evident that when a straight line is fitted through the data some points will be above it and some below, i.e. some of the residuals will be positive and some negative. The aim of all sorts of straight-line fitting is to get some form of 'balance' between these two types of residual and to ensure that when the residuals are plotted they are as patternless as possible. The method of line fitting that we will use, the least-squares method, attempts to remove any pattern from the residuals by minimizing the sum of the squared residuals. Before turning our attention to this method let us look briefly at some patterns of residuals from a fitted line.

Consider the following scattergrams (figures 7.5–7.7) of a small number of points and underneath them their residuals from the fitted line. These scattergrams and residual plots are illustrative and diagrammatic only and no scales have been given. However, the residuals under each scattergram should be seen as being plotted on a larger scale than the scattergram from which they are derived. Consider three general cases.

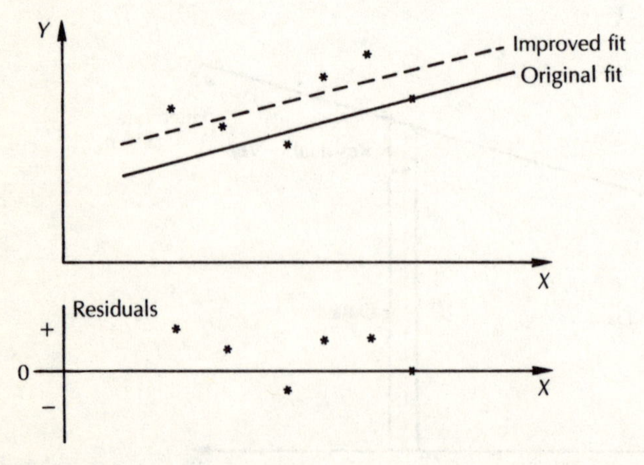

Figure 7.5

In figure 7.5 all residuals except one are positive or zero and the fit can probably be improved by moving the line upwards and parallel to itself as indicated.

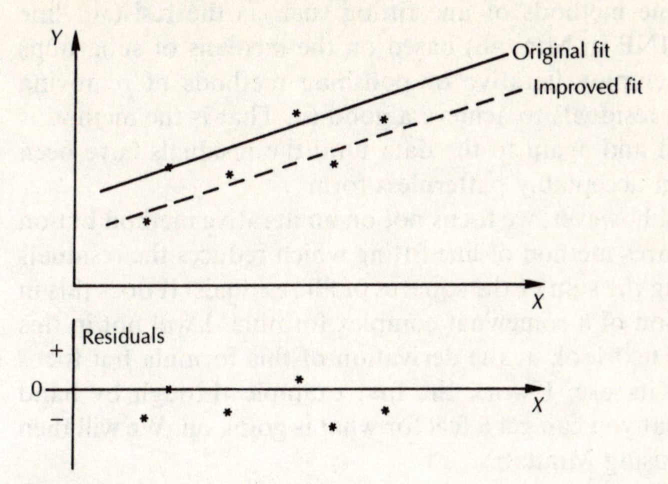

Figure 7.6

In figure 7.6 the fit could probably be improved by moving the original line down and parallel to itself as indicated.

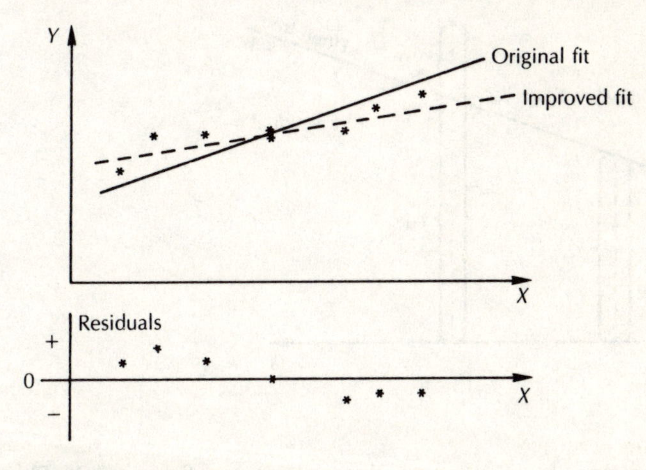

Figure 7.7

In figure 7.7 the residuals show a distinct clockwise tilt pattern. An improved fit line could probably be achieved by tilting the original fit line in a clockwise direction.

A consideration of residuals is important in achieving a good fit and the aim of any method of line fitting is, as we have mentioned, to try to achieve a scatter of residuals which is as patternless as possible. Some methods of line fitting such as the resistant line method (RLINE in Minitab) based on the medians of subgroups of the data employ iterative or polishing methods of removing pattern from residuals to achieve a good fit. That is the method is applied again and again to the data until the residuals have been reduced to an acceptably patternless form.

In this text, however, we focus not on an iterative method but on the least-squares method of line fitting which reduces the residuals by minimizing the sum of the squares of all residuals. It does this in one application of a somewhat complex formula. I will not in this introductory text look at the derivation of this formula but focus attention on its use. I work the first example through by hand initially so that you can get a feel for what is going on. We will then continue by using Minitab.

The Least-squares Regression Line

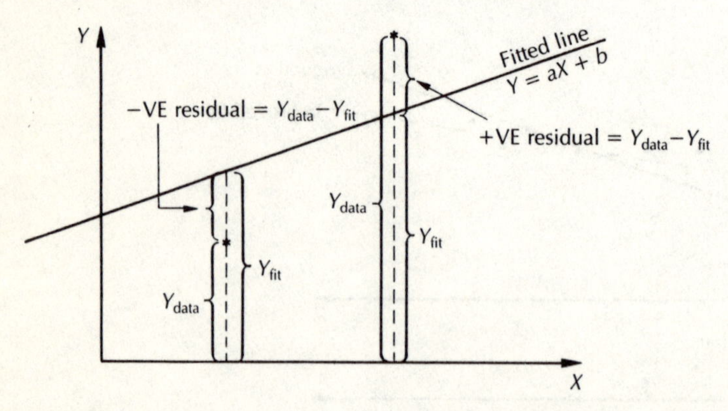

Figure 7.8

Consider the residuals of the two data points in figure 7.8. The residual for each data point is given by $Y_{\text{data}} - Y_{\text{fit}}$. It is our general aim in line drawing to minimize the sum of these residuals. However, as some residuals are positive and others are negative we cannot take a simple algebraic sum as these would tend to cancel each other out. What this method does is square each residual $Y_{data} - Y_{fit}$ and look for a line which minimizes this sum. Bear in mind that minus times minus gives a positive quantity. The least-squares regression line is the fitted line which minimizes the sum of the squared residuals, i.e.

$$\sum(Y_{data} - Y_{fit})^2 = \sum(\text{residual})^2$$

is minimized. As before we let the fitted line be $Y = aX + b$ where a is the slope or gradient and b is the intercept on the Y axis. Using the assumptions of the least-squares regression line method it can be shown that the slope a is given by the expression,

$$\text{slope} = a = \frac{N\sum(XY) - (\sum X)(\sum Y)}{N\sum X^2 - (\sum X)^2}$$

where X is the independent variable, Y is he dependent or response variable and N is the number of data points (x, y) in the set. This line $Y = aX + b$ passes through the point $(\bar{X}\ \bar{Y})$, the centre of gravity of the data set, and so we can find the intercept b by substituting in our general equation to get

$$\bar{Y} = a\bar{X} + b$$

i.e.

$$b = \bar{Y} - a\bar{X}$$

$\bar{X}$ and $\bar{Y}$ are the means of the X and Y values respectively and can be found from our previous formula for means:

$$\frac{\sum X}{N} = \bar{X} \text{ and } \frac{\sum Y}{N} = \bar{Y}$$

Let us remind ourselves what the terms in the equation of the slope mean:

$\sum X$ is the sum of the X values
$\sum Y$ is the sum of the Y values
$\sum(XY)$ is the sum of the cross products obtained by multiplying each X by its corresponding Y value and summing
$\sum X^2$ indicates that we multiply each X value by itself and take the sum
$(\sum X)^2$ is the sum for the X values multiplied by itself

Before looking at an example it is well to note that this method assumes that there is some degree of linear relation between the two variables and it is always wise to plot and look at the data before trying to fit a regression line.

We should also note that, unlike some methods of line fitting such as the resistant line fit (RLINE in Minitab), the procedure uses all the data and gives each point equal consideration. This has both advantages and disadvantages. All the data contribute to the plot but as the method uses squared residuals the effect of freak values or outliers makes it much less resistant. That is, odd values at the ends of the distribution may, when squared, exert a great leverage or moment on the results. The regression line goes through the 'middle' or centre of gravity of the data set as indicated by $(\bar{X}, \bar{Y})$ and the outliers or odd values which are far removed from this point may have a considerable rotating effect on the line about its middle point.

When we finish our analysis we can, should we so wish, plot and examine the residuals to evaluate how good the fit is. An additional bonus of the regression method, which we will not examine here, however, is that we can get various indices or measures of the strength of fit of the line from the residuals and the print-out provided by Minitab.

In order that the calculation by hand is not too cumbersome I have taken a sample of 10 from the data on seamen's heights and weights that we looked at in chapter 6. The data are given in table 7.1.

Table 7.1 Height and weight of ten seamen

No	*Name*	*Height X*	*Weight Y*
1	Coombe	177	62
2	Griffiths	172	55
3	Mitchell	188	80
4	Phipps	170	51
5	Youngman	182	71
6	Burman	178	66
7	Hawkins	178	68
8	Hicks	187	75
9	Maund	173	60
10	Baker	188	85

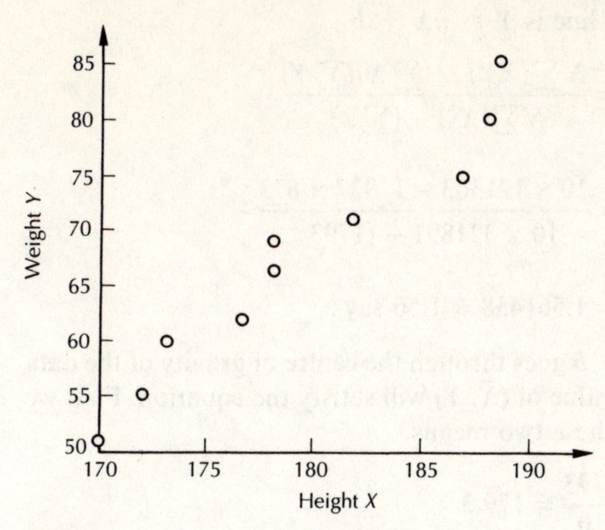

Figure 7.9

As we know from our previous work there is a high positive correlation between the variables of height and weight. Figure 7.9 reveals a fairly high degree of linearity in the data and we should be successful in fitting a straight line to the data.

In our analysis by hand the next step is to lay out the data in a tabular form (table 7.2), calculate the column entries and find the gradient of the line. As you can see I have also calculated the residuals in order to illustrate how they are obtained. However, our focus in this least-squares method will be on the elements needed for the gradient equation of the line $Y = aX + b$.

Table 7.2 Data for least-squares regression line

	X	Y_{data}	XY	X^2	Y_{fit}	*Residual* $(Y_{data} - Y_{fit})$
	177	62	10974	31329	63.72	−1.72
	172	55	9460	29584	55.92	−0.92
	188	80	15040	35344	80.88	−0.88
	170	51	8670	28900	52.80	−1.8
	182	71	12922	33124	71.52	−0.52
	178	66	11748	31684	65.28	0.72
	178	68	12104	31684	65.28	2.72
	187	75	14025	34969	79.32	−4.32
	173	60	10380	29929	57.48	2.52
	188	85	15980	35344	80.88	4.12
Totals	1793	673	121303	321891		

$N = 10$ and the line is $Y = aX + b$.

$$\text{gradient} = a = \frac{N\sum(XY) - (\sum X)(\sum Y)}{N\sum(X^2) - (\sum x)^2}$$

$$= \frac{10 \times 121303 - 17937 \times 673}{10 \times 321891 - (1793)^2}$$

$$= 1.561438 = 1.56 \text{ say}$$

The line $Y = aX + b$ goes through the centre of gravity of the data $(\bar{X}, \bar{Y})$ and so the value of $(\bar{X}, \bar{Y})$ will satisfy the equation. First we need to calculate these two means.

$$\bar{X} = \frac{\sum X}{N} = \frac{1793}{10} = 179.3$$

$$\bar{Y} = \frac{\sum Y}{N} = \frac{673}{10} = 67.3$$

The line goes through the centre of gravity of the data, the point (179.3, 67.3). Substituting these values in the equation $Y = aX + b$ we can find b, the intercept on the Y axis.

$$67.3 = 1.56 \times 179.3 + b$$

and so

$$b = 67.3 - (1.56 \times 179.3) = -212.4$$

The equation of the fitted line using the method of least squares is therefore

$$Y = 1.56X - 212.4$$

That is, for this fitted line the variables of height and weight are related to each other as follows:

$$\text{weight} = 1.56 \times \text{Height} - 212.4$$

(weight in kilograms, height in centimetres). I will now demonstrate how to calculate the residuals $Y_{data} - Y_{fit}$. To do this we need to find the Y_{fit} corresponding to each Y_{data}, i.e. put each X value in turn into the fit equation $Y_{fit} = 1.56X - 212.4$.

When X is 177 we have

$$Y_{fit} = 1.56 \times 177 - 212.4 = 63.72$$

$$\text{first residual} = Y_{data} - Y_{fit} = 62 - 63.72 = -1.72$$

When X is 172 we have

$$Y_{fit} = 1.56 \times 172 - 212.4 = 55.92$$

$$\text{second residual} = Y_{data} - Y_{fit} = 55 - 55.92 = -0.92$$

The other eight residuals are calculated in a similar fashion.

Let us now calculate the least-squares regression line for the data using Minitab. I will first reproduce a program for this work and the resulting analysis. I have embedded some comment in the print-out but I give a more extended account of the analysis later.

Least-squares Regression Line Using Minitab

To gain the maximum benefit from this and other exercises you should work it through for yourself while using the print-out as a guide. Don't forget to save your analysis with an OUTFILE command such as MTB > OUTFILE 'A:\LFITS01'.

```
MTB > READ the following data into C1 and C2
DATA> 177 62
DATA> 172 55
DATA> 188 80
DATA> 170 51
DATA> 182 71
DATA> 178 66
DATA> 178 68
DATA> 187 75
DATA> 173 60
DATA> 188 85
DATA> END
     10 ROWS READ
```

```
MTB > #Go into the editor and name the columns
MTB > #C1 X(Ht) C2 Y(Wt)

MTB > PLOT C2 against C1
```

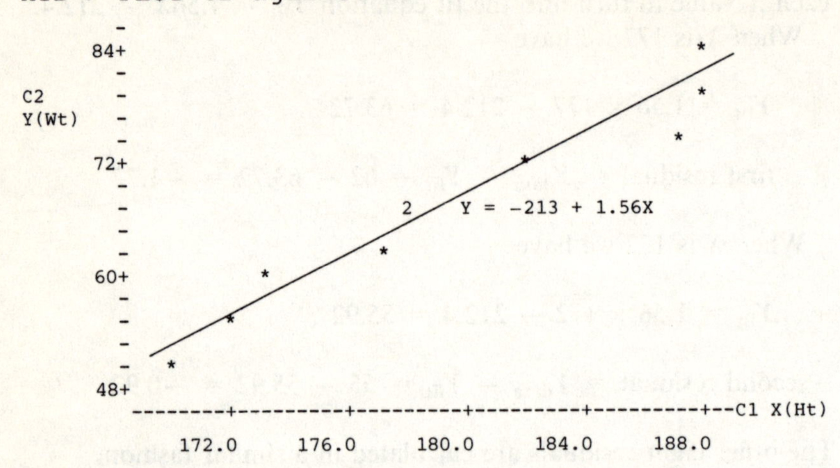


```
MTB > #The BRIEF command controls the detail and
MTB > #amount of output. The REGRESS command
MTB > #gives the least-squares regression
MTB > #analysis.
MTB > #The subcommand stores the Residuals.
MTB > BRIEF=3
MTB > REGRESS the Y values in C2 on 1 predictor
MTB > #in C1;
SUBC> RESIDUALS put in C3.

The regression equation is
Y(Wt) = -213 + 1.56X(Ht)
Predictor      Coef      Stdev    t-ratio       p
Constant    -212.67      23.97      -8.87   0.000
X(Ht)        1.5614     0.1336      11.69   0.000

s = 2.692    R-sq = 94.5%      R-sq(adj) = 93.8%

Analysis of Variance

SOURCE        DF        SS        MS        F       p
Regression     1    990.11    990.11   136.59   0.000
Error          8     57.99      7.25

Total          9   1048.10
```

```
Obs.  C1       C2      Fit   Stdev.Fit   Residual   St.Resid
  1  177   62.000   63.709       0.905     -1.709      -0.67
  2  172   55.000   55.902       1.295     -0.902      -0.38
  3  188   80.000   80.885       1.441     -0.885      -0.39
  4  170   51.000   52.779       1.506     -1.779      -0.80
  5  182   71.000   71.516       0.925     -0.516      -0.20
  6  178   66.000   65.270       0.869      0.730       0.29
  7  178   68.000   65.270       0.869      2.730       1.07
  8  187   75.000   79.323       1.335     -4.323      -1.85
  9  173   60.000   57.463       1.197      2.537       1.05
 10  188   85.000   80.885       1.441      4.115       1.81

MTB > #Go into the editor and name C3 Residual
MTB > SET C4
DATA> 1:10
DATA> END
MTB > #The above puts ten numbers into C4
MTB > #and we can use these to plot the Residuals.
MTB > PLOT C3 against C4
```

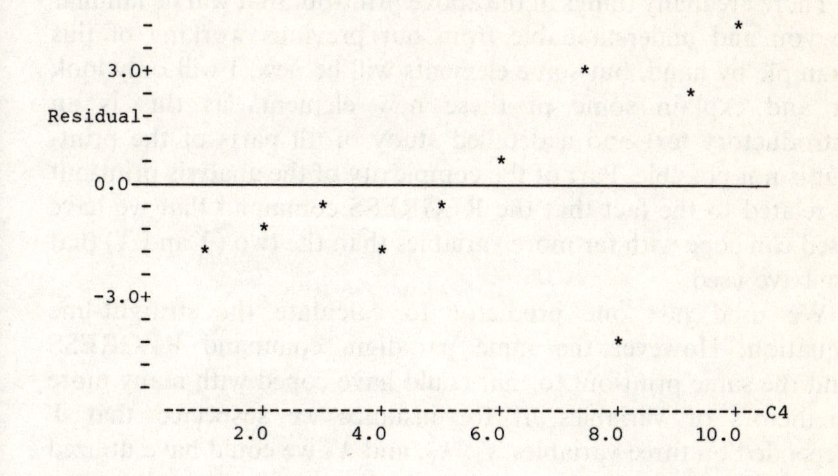


```
MTB > #This type of plot of the residuals may be
MTB > #of use in some cases but generally it is
MTB > #more instructive to plot them against the X
MTB > #(explanatory) variable as follows.
```

```
MTB > PLOT C3 against C1
```

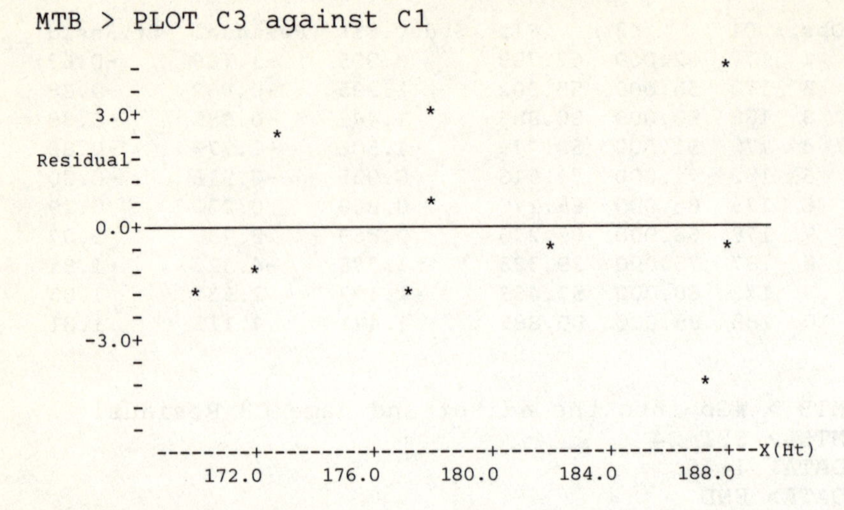


```
MTB > #These residuals appear to be relatively
MTB > #patternless.
```

There are many things in the above print-out that will be familiar to you and understandable from our previous working of this example by hand, but some elements will be new. I will only look at and explain some of these new elements as this is an introductory text and a detailed study of all parts of the print-out is not possible. Part of the complexity of the analysis print-out is related to the fact that the REGRESS command that we have used can cope with far more variables than the two (Y and X) that we have used.

We used just one predictor to calculate the straight-line equation. However the same paradigm command REGRESS and the same print-out format could have coped with many more predictors or variables. If for instance we suspected that Y depended on three variables X_1, X_2, and X_3 we could have utilized a model such as, say,

$$Y = a_1X_1 + a_2X_2 + a_3X_3 + b$$

However, to get back to our simple two-variable model, there are some elements of the print-out, such as the R^2 value, that merit our attention. First let me say a few words about the form of the initial BRIEF and REGRESS commands that initiate the analysis.

The BRIEF command controls the amount of output from the REGRESS command analysis. There are four options: BRIEF = 0, BRIEF = 1, BRIEF = 2 and BRIEF = 3. The first gives no output (!), but any storage of residuals etc. is done. BRIEF = 1 gives the least-squares regression equation, R^2 values and a small amount of output. The command BRIEF = 3 that I have used above gives the full output with equation, various factors, residuals and fits etc. BRIEF = 2 gives an intermediate amount of output including the equation, R^2 and an analysis of variance table and notes the outliers or unusual observations in the table of fits and residuals. For most of our purposes in this text the command BRIEF = 2 is probably sufficient (see the next example).

The REGRESS command can be used with or without subcommands. Remember the punctuation at the end of each line if you do use a subcommand. As you can see, in the above analysis I have used a subcommand to get the residuals stored in C3. They can be examined using the HISTOGRAM command or, as I have done, by SETting another column with the same number of values, in our case the numbers 1 2 3 ... 10, and then plotting. In general it is more useful to plot the residuals against the X or explanatory variable. However, if we transform the X variable, as in some examples in the next chapter, we may find it useful to plot the residuals against the transformed X variable.

Our plotted residuals do not show any distinctive pattern about the zero line that I have drawn in. It should be noted that we are more likely to be interested in the pattern of the residuals if we had employed an iterative method, such as RLINE, to repeatedly examine and polish our results to achieve a better and better fitted line. However even with the least-squares method that we have used a plot of the residuals is usually desirable and can be a very useful way of revealing any serious departures from the straight-line model.

The first element of the printed analysis is the least-squares regression equation which is our main focus of interest in this chapter. It is given as

$$Y(\text{Wt}) = -213 + 1.56X(\text{Ht})$$

which is of course very close to our equation calculated previously by hand, the small difference being due to number rounding.

Of the other elements in the print-out we need, at this stage, take note only of the quantity called R^2. This is sometimes called the coefficient of determination and the larger its value the better the model. That is, R^2 is a measure of how successfully the straight line $Y = aX + b$ has been fitted to the empirical data set by the least-squares method employed. In our case the value of R^2 at (94.5%) is very high indeed. This means that the line $Y = -213 + 1.56X$ is an extremely good fit for the data. This can be seen from the line that I have drawn into the first of the figures in the print-out. This has had to be done by hand as Minitab unfortunately cannot draw the graphs of given equations.

In a least-squares regression analysis with one explanatory variable (one predictor) such as we have here, the R^2 (coefficient of determination) is equal to the square of the coefficient of correlation that we examined in the last chapter. However, we should note that the coefficient of correlation involved here is the Pearson product moment correlation rather than Spearman's rank correlation coefficient.

We may think of R^2 as being a measure of the amount of variability in the empirical data that is 'explained' or measured by the fitted line. This can be expressed as

$$R^2 = \frac{\text{sum of the squares explained by the straight line model}}{\text{total sum of the squares}}$$

$$= \frac{SS}{\text{total } SS}$$

If you look at the entries in the analysis of variance part of the print-out you will see that the ratio of SS/total SS is 990.11/1048.10 which is equal to 94.5%, i.e. R^2, the coefficient of determination. The value of R^2 is very high and usually we will need to be satisfied with lines which fit the data less well.

Let us now look at another example of line fitting using the least-squares method with a much larger data set. We consider the part of the file DATA11.DAT that records the infant and maternal mortality rates (IMR and MMR).

In Chapter 6, Exercise 5, I asked you to plot IMR against MMR, examine the plot and decide whether the data were sufficiently linear in form to justify correlating the two variables.

There did appear to be a fair amount of linearity and for the purposes of the present exercise we will assume that there is enough linearity both to calculate a correlation coefficient and to see if we can connect the two variables with a straight-line equation. Please work this example through for yourself and use the print-out as a guide.

As before I have embedded a few comments in the Minitab print-out but I will give a more extensive commentary later.

```
MTB > READ 'A:\DATA11.DAT' into C1-C12
   59 ROWS READ
ROW   C1      C2   C3   C4   C5   C6   C7   C8   C9   C10
  1    1      80    5   10   35   17  173   45   21   300
  2    1     610    6   15   34    *  173   56   29     *
  3    1     180    7   11    7   17  144    *    *   170
  4    1     120    *    *    *   34  130    *    *     *
 . . .
ROW    C11     C12

  1     48    15.7
  2     46    10.0
  3     48     8.8
  4     46    49.2
 . . .
MTB > #Go into the Editor and name C7 Y(IMR)
MTB > #and C10 X(MMR).

MTB > PLOT C7 against C10
```

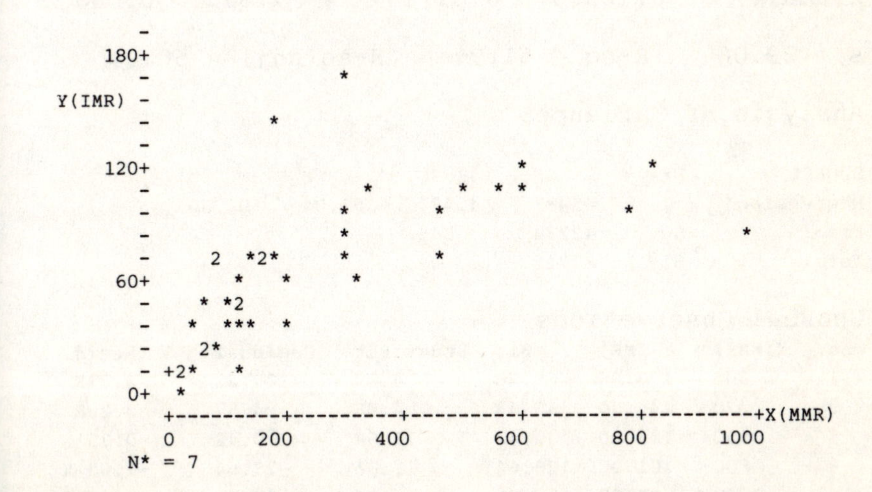

```
MTB > DESCRIBE C7 C10

              N       N*     MEAN    MEDIAN    TRMEAN
Y(IMR)       59        0    55.03     48.00     51.68
X(MMR)       52        7    199.1     120.0     167.8

          STDEV   SEMEAN      MIN       MAX        Q1        Q3
Y(IMR)    43.88     5.71     5.00    173.00     11.00     82.00
X(MMR)    237.5     32.9      2.0    1000.0      13.5     300.0

MTB > #The MMR has a very high standard deviation.
MTB > #Lots of variability here.
MTB > CORRELATE C7 and C10

Correlation of Y(IMR) and X(MMR) = 0.720

MTB > # This time we will use BRIEF=2 but again we
MTB > # will collect the residuals so we need a
MTB > # subcommand.
MTB > BRIEF=2
MTB > REGRESS Y values in C7 on 1 predictor in C10;
SUBC> RESIDUALS into C13.

The regression equation is
Y(IMR) = 28.1 + 0.126 X(MMR)

52 cases used 7 cases contain missing values

Predictor        Coef      Stdev    t-ratio        p
Constant       28.059      5.281       5.31    0.000
X(MMR)        0.12573    0.01713       7.34    0.000

s = 29.06     R-sq = 51.9%     R-sq(adj) = 50.9%

Analysis of Variance

SOURCE        DF        SS        MS         F         p
Regression     1     45481     45481     53.84     0.000
Error         50     42234       845
Total         51     87715

Unusual Observations
Obs.  X(MMR)    Y(IMR)      Fit  Stdev.Fit   Residual   St.Resid
  1      300    173.00    65.78       4.39     107.22      3.73R
  3      170    144.00    49.43       4.06      94.57      3.29R
  5      830    123.00   132.42      11.54      -9.42     -0.35 X
  9      800    101.00   128.64      11.06     -27.64     -1.03 X
 14     1000     86.00   153.79      14.30     -67.79     -2.68RX
```

```
R denotes an obs. with a large st. resid.
X denotes an obs. whose X value gives it
large influence.

MTB > #These marked observations will need
MTB > #our attention.
MTB > #Let us plot the residuals.
MTB > PLOT C13 against C10
```

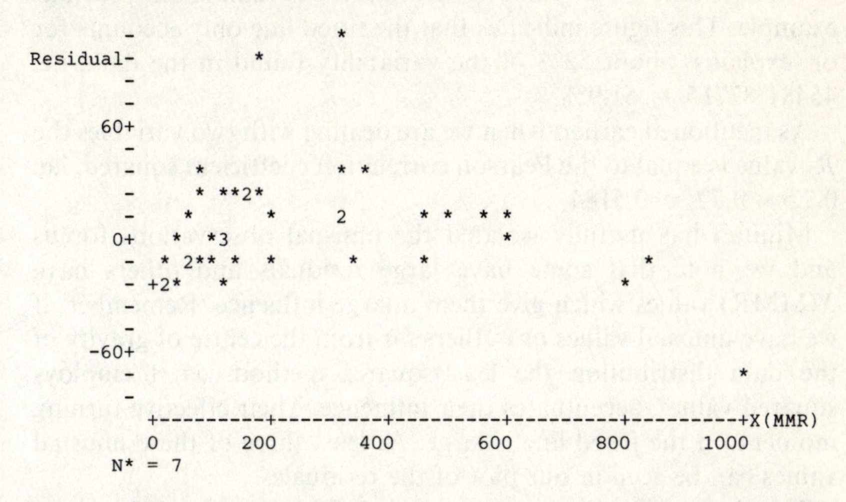

```
MTB > #These residuals are not as patternless as
MTB > #we would like. Probable indication of some
MTB > #curvilinear pattern in the original data
MTB > #that we missed.
```

The greater mass of the data appears to be contained within a lozenge-shaped pattern spreading from the lower left-hand corner of the graph to the upper right. We can probably fit a straight line to it. However, there are a few odd values or outliers which need to be looked at. It is probably worth our while to collect the residuals from the fit and then examine them.

The statistical description indicates that there is a great deal of variability amongst the maternal mortality rates as measured by the standard deviation.

We note that the correlation coefficient (Pearson's product moment) is 0.72.

In this analysis I have used BRIEF = 2 to control the amount of output; if more output was required BRIEF = 3 could have been used. I have collected the residuals again so that we can examine them and identify the outliers or unusual values.

The least-squares regression equation calculated from the data is

$$Y(\text{IMR}) = 28.1 + 0.126X(\text{MMR})$$

We note that the R^2 value at 51.9% is lower than in our previous example. This figure indicates that the fitted line only accounts for or 'explains' about 52% of the variability found in the data, i.e. 45481/87715 = 51.9%.

As mentioned earlier, when we are dealing with two variables the R^2 value is equal to the Pearson correlation coefficient squared, i.e. $0.72 \times 0.72 = 0.5184$.

Minitab has usefully isolated the unusual observations for us and we note that some have large residuals and others have X(MMR) values which give them a large influence. Remember, if we have unusual values or outliers far from the centre of gravity of the data distribution the least-squares method, as it employs squared values, accentuates their influence. Their effective turning moment on the fitted line is large. At least three of these unusual values can be seen in our plot of the residuals.

The data points that are unusual and need our attention are from the following sources (see the copy of the file DATA11.DAT in appendix 1).

1 Mozambique
3 Malawi
5 Nepal
9 Nigeria
14 Ghana

If we were doing a research project rather than an illustrative exercise we would need to check the data at this point for each of these countries. Are the data correct? Has a transcription error been made? The maternal mortality rate of 1000 for Ghana seems to be improbably high. However, a check with the UNICEF data indicates that no simple transcription error has been made.

At this point it is worth noting that all the above mentioned countries are in the economically developing world and that the standard of both health care and the collection of vital data, especially in rural areas, is likely to be very poor.

In addition we could also note that in recent years as computer databases and analysis packages have become more widely available and powerful the amount of data collected and processed has greatly increased. Some 'official statistics' quoted by such well established bodies as UNICEF and other United Nations agencies as well as by organizations such as the World Bank often need to be treated with caution. As Durkheim mentioned a long time ago in relation to suicide rates, we need to ask whether the statistics on which we base our analyses reflect the true incidence of the variable that we are trying to measure and assess (suicide, infant or maternal mortality etc.) or the vagaries of the way the data were collected.

However, to return to the mortality data and our attempt to capture the relationship between the two variables in a neat straight-line equation, we can say that our efforts have been reasonably successful, but we should bear in mind that the 59 countries considered vary enormously in health care, wealth and other characteristics. The type of analysis that we have done here would no doubt be greatly improved if we considered the two variables IMR and MMR in relation to subgroups of a set of countries. The countries could be divided into subgroups by, for instance, level of gross national product (GNP) or according to the under 5s mortality rate (U5MR). Also, as mentioned previously, it is likely that both the indices of mortality that we have used in this simple illustrative analysis are connected to a third factor, the amount spent on health care in each country.

Let us now look back to the plot of the residuals from the fitted line. We can see that when they are plotted against the explanatory variable X, a pattern emerges. They seem to lie in a arc-like shape about the zero line. This is an indication that there was a curvilinear pattern in the original data, i.e. in the IMR and MMR data, that was not evident to us from our initial plot. In view of this pattern in the residuals it would probably be wise to transform the original data before attempting to fit a straight line. We look at such transformations in the next chapter.

To end this chapter let us try to fit a straight line to some more of the UNICEF data in file DATA11.DAT.

Let us investigate the relation between infant mortality and female literacy. These variables are contained in columns C7 and C9 of the data file (see appendix 1). As before you should work this example through for yourself using the print-out as a guide.

```
MTB > READ 'A:\DATA11.DAT' into C1-C12
   59 ROWS READ
ROW  C1    C2  C3    C4   C5   C6    C7   C8   C9   C10
  1   1    80   5    10   35   17   173   45   21   300
  2   1   610   6    15   34    *   173   56   29     *
  3   1   180   7    11    7   17   144    *    *   170
  4   1   120   *     *    *   34   130    *    *     *
.   .   .
ROW   C11    C12

  1    48   15.7
  2    46   10.0
  3    48    8.8
  4    46   49.2
.    .    .
MTB > #Go into the Editor and name C7 Y(IMR) and
MTB > #C9 X(FELIT).
MTB > PLOT C7 against C9
```

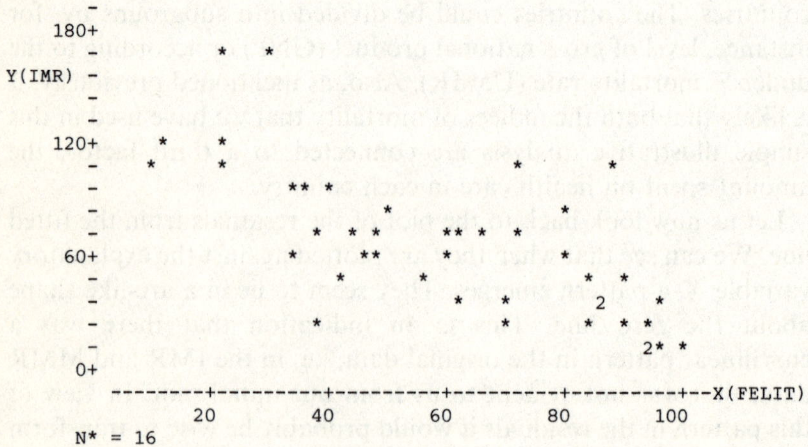

```
MTB > CORRELATE C7 and C9

Correlation of Y(IMR) and X(FELIT) = -0.720

MTB > #Let us now get the least-squares regression
MTB > #as there does appear to be a fair degree of
MTB > #linearity in the above data. We will use
MTB > #BRIEF=2 and also collect the residuals.
MTB > BRIEF=2
MTB > REGRESS Y values in C7 on 1 predictor in C9;
SUBC> RESIDUALS into C13.

The regression equation is
Y(IMR) = 132 - 1.10 X(FELIT)

43 cases used 16 cases contain missing values

Predictor        Coef       Stdev     t-ratio        p
Constant       131.88       10.92       12.08    0.000
X(FELIT)      -1.1032      0.1661       -6.64    0.000

s = 27.71      R-sq = 51.8%      R-sq(adj) = 50.6%

Analysis of Variance

SOURCE          DF         SS          MS        F        p
Regression       1      33856       33856    44.08    0.000
Error           41      31488         768
Total           42      65344


Unusual Observations
Obs.X(FELIT)  Y(IMR)      Fit  Stdev.Fit  Residual  St.Resid
  1     21.0  173.00   108.71       7.82     64.29     2.42R
  2     29.0  173.00    99.89       6.74     73.11     2.72R
  8     88.0  102.00    34.80       6.21     67.20     2.49R
 40     38.0   24.00    89.96       5.65    -65.96    -2.43R

R denotes an obs. with a large st. resid.

MTB > #The marked observations will require
MTB > #our attention.
MTB > #Let us now plot the residuals.
MTB > HISTOGRAM C13
```

```
Histogram of C13 N = 43 N* = 16

Midpoint  Count
     -60      1 *
     -40      4 ****
     -20     11 ***********
       0     16 ****************
      20      6 ******
      40      2 **
      60      2 **
      80      1 *

MTB > #Go into the editor and label C13 Residual.
MTB > #We also need to SET a column with 59
MTB > #numbers to plot the Residuals against.
MTB > SET C15
DATA> 1:59
DATA> END
MTB > PLOT C13 against C15
```

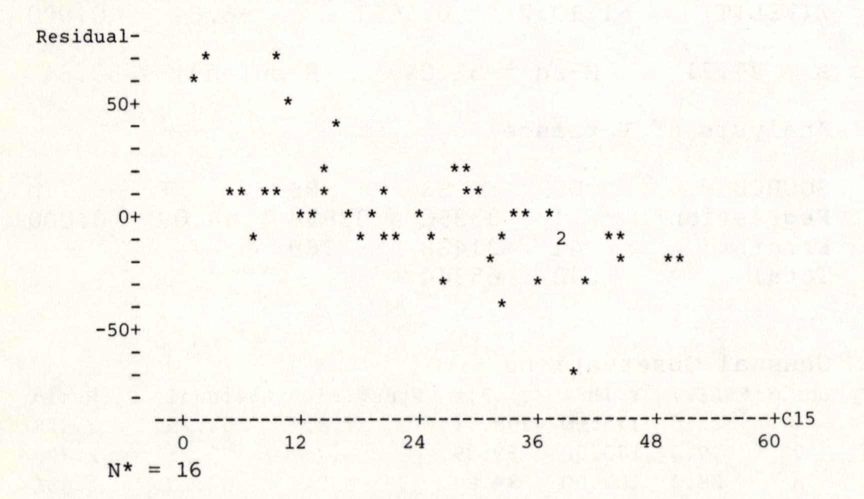


```
MTB > #Let us plot the Residuals against the
MTB > #explanatory variable of female literacy.
MTB > PLOT C13 against C9
```

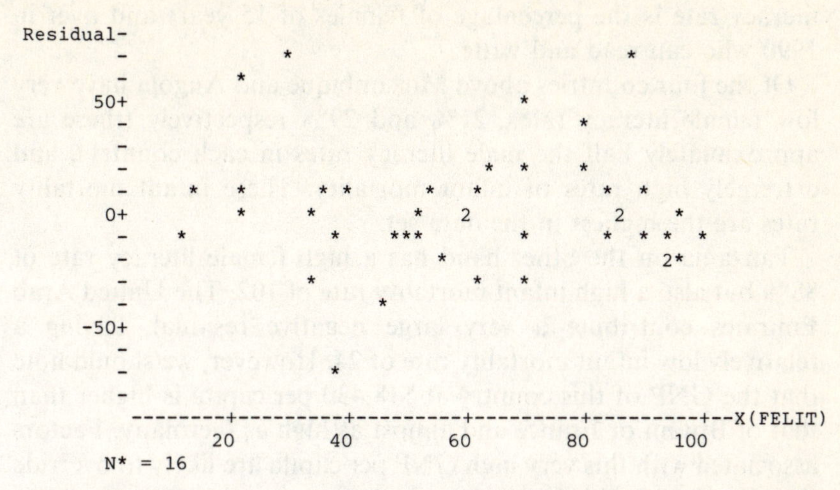

We note from the plot of the bivariate data that there seems to be a high degree of linearity. The correlation of infant mortality and female literacy is a fairly high negative figure of −0.72. We would expect from the appearance of the scattergram and this correlation that the least-squares regression line will give a reasonable fit to the data.

I have used BRIEF = 2 and again collected the residuals so that we can examine them.

The regression equation is $Y(\mathrm{IMR}) = 132 - 1.1X(\mathrm{FELIT})$ and as we expected from the value of the correlation coefficient R^2 is about 52%.

With this value of R^2 we have some large residuals and these are identified as occurring in countries 1, 2, 8 and 40. Referring to the file in appendix 1 we see that these are the following countries, and the values of the two variables are as shown:

		Female literacy	*Infant mortality rate*
1	Mozambique	21%	173
2	Angola	29%	173
8	Tanzania	88%	102
40	United Arab Emirates	38%	24

Let us remind ourselves what the variables measure. The infant mortality rate is the annual number of deaths of infants under 1 year of age per 1000 live births in the year 1990. The adult female literacy rate is the percentage of females of 15 years and over in 1990 who can read and write.

Of the four countries above Mozambique and Angola have very low female literacy rates, 21% and 29% respectively (these are approximately half the male literacy rates in each country), and extremely high rates of infant mortality. These infant mortality rates are the highest in the data set.

Tanzania on the other hand has a high female literacy rate of 88% but also a high infant mortality rate of 102. The United Arab Emirates contribute a very large negative residual, having a relatively low infant mortality rate of 24. However, we should note that the GNP of this country at $18,430 per capita is higher than that of Britain or France and almost as high as Germany. Factors associated with this very high GNP per capita are likely to override the effects of female literacy.

These unusual observations that Minitab has drawn our attention to are both useful in themselves and also because they stimulate further questions about the type of relationship that exists between the two variables in different social and economic circumstances. For instance the case of the United Arab Emirates suggests that, given time, it would be interesting to examine how the two variables of female literacy and infant mortality relate to each other in the context of Islamic cultures. A brief glance at the subset of Islamic countries indicates some very low rates of female literacy, many being less than half the male rate (table 7.3).

Table 7.3 Literacy rate

	Male %	*Female %*
Bangladesh	47	22
Sudan	43	12
Pakistan	47	21
Libya	75	50
Morocco	61	38
Algeria	70	46
Saudi Arabia	73	48
Iraq	70	49

However, these interesting avenues of speculation and exploration unfortunately cannot be pursued in this introductory text and so we return to consider the residuals.

As you can see there seems to be no apparent pattern in the plotted residuals and with R^2 at 52% we can assume that the fitted line Y(IMR) = 132 – 1.1X(FELIT) is a useful summary of the relationship between these two variables in this data set.

The fairly close relationship between these two variables may surprise you but it is documented in other sources. For instance Professor J. Caldwell states, on the basis of a study in Nigeria, that 'Maternal education is the single most significant determinant of these marked differences in child mortality. Such differences are also affected by a large range of other socio-economic factors, but no other factor has the impact of maternal education' (Caldwell, 1981; quoted by Peter Adamson, 1991).

Postscript

In this chapter we have examined the equation of a straight line and have defined the terms fit and residual. We have looked at a very powerful method of fitting a straight line to suitable empirical data. Data which are roughly lenticular in form when plotted can be neatly summarized by a straight line calculated by the method of least squares.

In the next chapter we continue to employ the least-squares regression method of line fitting but we examine how it can be used with data which show some degree of curvilinear pattern. That is, we look at some simple types of transformation that can be used to 'straighten' data.

Exercises

1 Find the gradient, intercept and the equations of the straight lines through the (X, Y) points indicated:

(a) A(0,1) B(5,3)
(b) A(–1,0) B(6,6)
(c) A(1,8) B(12,0)

2 A researcher who is interested in predicting the weights of premature babies based on their age in weeks has collected the following data on eight premature infants given in table 7.4

Table 7.4

Infant	*Age in weeks (X)*	*Weight in kilograms (Y)*
PA	4	2.3
PB	4	1.8
PC	5	2.3
PD	1	1.0
PE	1	1.1
PF	2	1.2
PG	3	1.4
PH	6	2.5

(i) Calculate the least-squares regression line for the data by 'hand'.

(ii) Use Minitab to plot a scatter gram of the data and calculate the least-squares regression line.

(iii) What is the predicted weight of a three week old premature baby based on the above data?

3 The data in table 7.5 come from a study of 21 children with cyanotic heart disease. Two variables are given for each child, the first (*X*) being the age of the child in months when he\she uttered the first word, the second (*Y*) being the child's score on a scale of adaptation, the Gesell Adaptation test.

Plot the data and fit a straight line by the method of least squares. Collect the residuals and comment briefly on the fit.

4 In 1947 a large-scale survey of the intelligence of all 11-year old-Scottish school children was undertaken. An intelligence test was administered to 75,451 children and a comprehensive amount of socio-economic and health data was collected about each child.

(i) When the intelligence test scores were grouped by family size the data given in table 7.6 resulted.

Plot the data and calculate the least-squares regression line to fit the data. Comment briefly on your results.

Table 7.5

Child	*X*	*Y*
1	15	95
2	26	71
3	10	83
4	9	91
5	15	102
6	20	87
7	18	93
8	11	100
9	8	104
10	20	94
11	7	113
12	9	96
13	10	83
14	11	84
15	11	102
16	10	100
17	12	105
18	42	57
19	17	121
20	11	86
21	10	100

Source: Quoted in Kelly, 1992

Table 7.6

Size of family (X)	*Mean test score (Y)*
1	42.03
2	41.74
3	38.32
4	35.32
5	32.51
6	30.88
7	29.45
8	28.81
9	27.97
10	26.94
11	27.34
12 and over	24.26

Source: Maxwell 1953

(ii) A representative sample of 3404 boys was drawn from the population of 75,451 children and when their weight in pounds was measured and related to family size the following mean weights were as given in table 7.7.

Table 7.7

Size of Family (X)	*Mean Weight* (Y)
1	72.1
2	71.7
3	70.9
4	69.5
5	69.3
6	68.8
7	68.7
8	68.5
9 and over	67.8

Source: Maxwell, 1953

Plot the data and calculate the least-square regression line to fit the data. Comment briefly on your results.

Answers to Exercises

1 (a) $\text{gradient} = a = \dfrac{\text{change of } y}{\text{change of } x}$

$$= \frac{1-3}{0-5} = \frac{-2}{-5} = \frac{2}{5}$$

$Y = aX + b$ satisfies (0, 1) and so $1 = 2/5 \times 0 + b$ and $b = 1$. The equation is

$$Y = \frac{2}{5}X + 1 \qquad \text{or} \qquad 5Y = 2X + 5$$

(b) $\text{gradient} = a = \dfrac{0-6}{-1\ -6} = \dfrac{-6}{-7} = \dfrac{6}{7}$

$Y = aX + b$ satisfies $(-1, 0)$ and so $0 = 6/7 \times (-1) + b$ and $b = 6/7$. The equation is

$$Y = \frac{6}{7}X + \frac{6}{7} \qquad \text{or} \qquad 7Y = 6X + 6$$

(c) $\text{gradient} = a = \dfrac{8-0}{1-12} = \dfrac{8}{11} = \dfrac{-8}{11}$

$Y = aX + b$ satisfies (1, 8) and so $8 = (-8/11) \times 1 + b$ and $b = 8 + 8/11 = 96/11$. The equation is

$$Y = \frac{-8X}{11} + \frac{96}{11} \quad \text{or} \quad 11Y = -8X + 96$$

2 (i)

X	Y_{data}	XY	X^2
4	2.3	9.2	16
4	1.8	7.2	16
5	2.3	11.5	25
1	1.0	1.0	1
1	1.1	1.0	1
2	1.2	2.4	4
3	1.4	4.2	9
6	2.5	15.0	36

$$\sum X = 26 \quad \sum Y_{\text{data}} = 13.6 \quad \sum XY = 51.6 \quad \sum X^2 = 108$$

N=8 and the line is $Y = aX + b$. The gradient a=0.31489 and $X = 3.25$, $Y = 1.7$, to give $b = 0.6766$. Therefore the regression line is Y=0.31489X + 0.6766
i.e. Y=0.32X + 0.68

(ii)

```
MTB > #This is the premature baby data.
MTB > #Enter the data in the Editor and label the
MTB > #columns C1 Weeks(X) C2 Kilos(Y)
MTB > PRINT C1 C2

 ROW   Weeks(X)   Kilos(Y)
   1          4        2.3
   2          4        1.8
   3          5        2.3
   4          1        1.0
   5          1        1.1
   6          2        1.2
   7          3        1.4
   8          6        2.5
```

```
MTB > PLOT C2 against C1
```

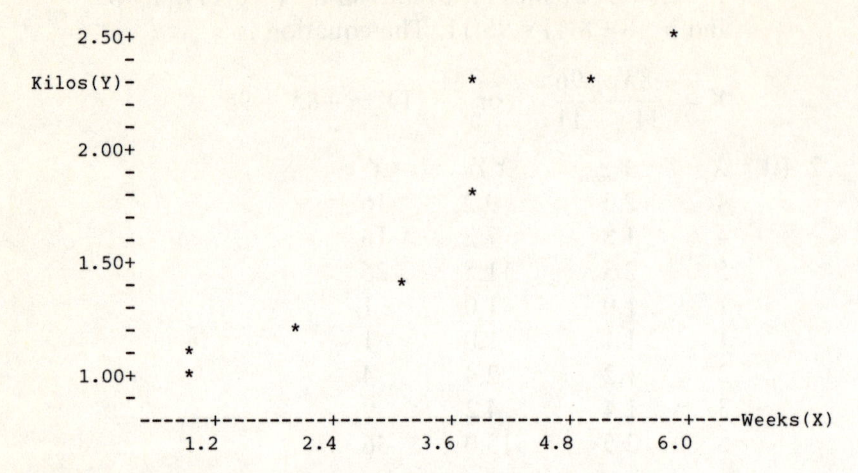


```
MTB > CORRELATE C1 and C2

Correlation of Weeks(X) and Kilos(Y) = 0.954

MTB > #Use the least-squares regression method
MTB > #and collect the residuals.
MTB > BRIEF=2
MTB > REGRESS Y values in C2 on 1 predictor in C1;
SUBC> RESIDUALS into C3.

The regression equation is
Kilos(Y) = 0.677 + 0.315 Weeks(X)

Predictor       Coef       Stdev    t-ratio        p
Constant      0.6766      0.1483       4.56    0.004
Weeks(X)     0.31489     0.04037       7.80    0.000

s = 0.1957    R-sq = 91.0%     R-sq(adj) = 89.5%

Analysis of Variance

SOURCE        DF        SS         MS         F        p
Regression     1    2.3302     2.3302     60.84    0.000
Error          6    0.2298     0.0383
Total          7    2.5600

Unusual Observations
Obs.Weeks(X) Kilos(Y)     Fit  Stdev.Fit  Residual  St.Resid
1 4.00         2.3000  1.9362     0.0755    0.3638     2.02R

R denotes an obs. with a large st. resid.
```

```
MTB > #Go into the Editor and name C3 Residual.
MTB > PLOT C3 against C1
```

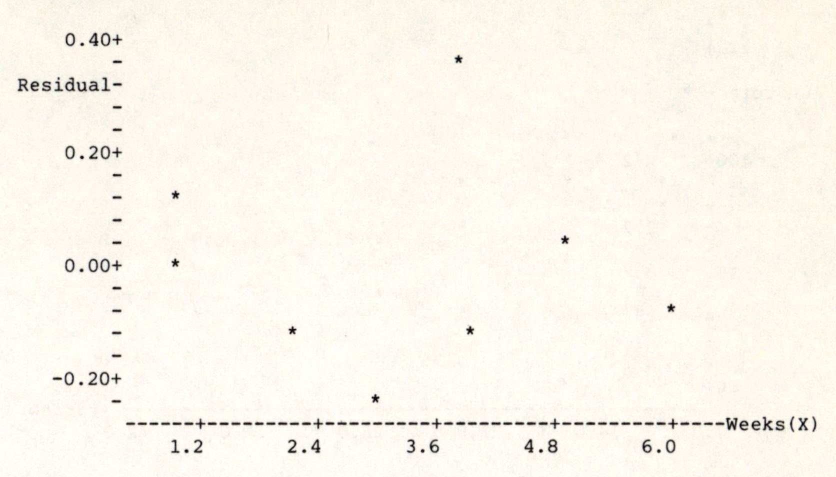

(iii) The fitted equation is kilos (Y) = 0.677 + 0.315 Weeks (X) and so putting in the X value of 3 weeks we get a predicted weight of

$$\text{kilos } (Y) = 0.677 + 0.315 \times 3 = 1.62 \text{ Kg}$$

3

```
MTB > #This is the data on cyanosed babies and
MTB > #adaptation.
MTB > #Enter the data into the Editor and name the
MTB > #columns
MTB > #C1 FirstWrd C2 GaScore.
MTB > SAVE 'A:\FirstWrd'

Worksheet saved into file: A:\FirstWrd.MTW
MTB > #The above saves the entered data for
MTB > #later use.
```

```
MTB > PLOT C2 against C1
```

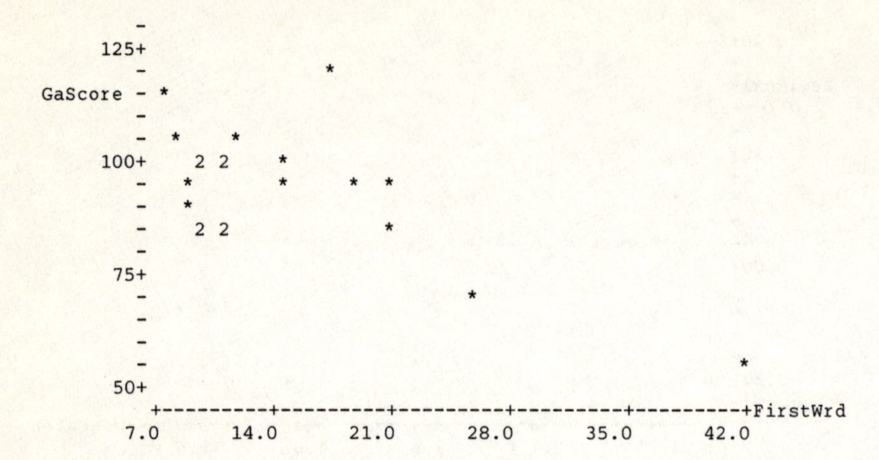


```
MTB > CORRELATE C2 and C1
```

Correlation of GaScore and FirstWrd = -0.640

```
MTB > BRIEF=2
MTB > REGRESS Y values in C2 on 1 predictor in C1;
SUBC> RESIDUALS into C3.
```

The regression equation is
GaScore = 110 - 1.13 FirstWrd

Predictor	Coef	Stdev	t-ratio	p
Constant	109.874	5.068	21.68	0.000
FirstWrd	-1.1270	0.3102	-3.63	0.002

s = 11.02   R-sq = 41.0%   R-sq(adj) = 37.9%

Analysis of Variance

SOURCE	DF	SS	MS	F	p
Regression	1	1604.1	1604.1	13.20	0.002
Error	19	2308.6	121.5		
Total	20	3912.7			

Unusual Observations

Obs.	FirstWrd	GaScore	Fit	Stdev.Fit	Residual	St.Resid
18	42.0	57.00	62.54	8.90	-5.54	-0.85 X
19	17.0	121.00	90.72	2.54	30.28	2.82R

```
R denotes an obs. with a large st. resid.
X denotes an obs. whose X value gives it large
  influence.

MTB > #Go into Editor and label C3 Residual.
MTB > PLOT C3 against C1
```

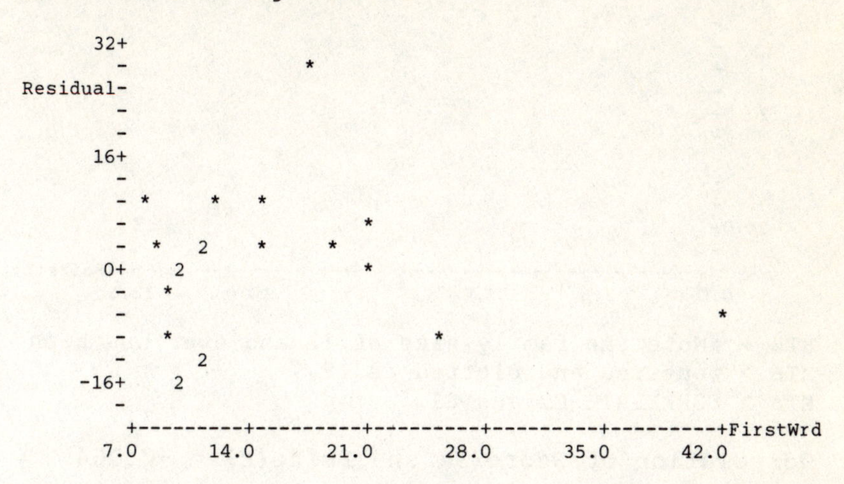

The fitted line has an R^2 of only 41% and we note that one of the residuals is extremely large (child 19 with an adaptation score of 121). The X value of Child 18 also gives it a large influence and residual.

When we plot the residuals we note that there may be a curvilinear pattern in them. In view of this we might have obtained a better fit if we had calculated a more resistant type of line, i.e. one not so susceptible to the influence of outliers. Minitab's RLINE is such a fit; otherwise we could have tried transforming the data before fitting.

4 (i)

```
MTB > #This is the data from the 1947 Scottish
MTB > #Mental Survey regarding the size of
MTB > #families and mean test scores.
MTB > #Enter the data into the Editor and name the
MTB > #columns C1 Fsize(X) and C2 Score(Y)
MTB > SAVE 'A:\SCOTS01'
Worksheet saved into file: A:\SCOTS01.MTW
MTB > #The above command saves the data.
```

```
MTB > PLOT C2 against C1
   42.0+       *    *
       -
Score(Y)-
       -              *
       -
   36.0+
       -                         *
       -
       -                                *
       -                                       *
   30.0+                                              *
       -                                                     *
       -                                                            *         *
       -                                                                   *
       -
   24.0+                                                                          *
       -
        +---------+---------+---------+---------+---------+Fsize(X)
       0.0       2.5       5.0       7.5      10.0      12.5
```

```
MTB > #Note the family size of 12 and over has been
MTB > #entered and plotted as 12.
MTB > CORRELATE C2 and C1

Correlation of Score(Y) and Fsize(X) = -0.964

MTB > #We will now regress Y on X using the
MTB > #least-square method and collect the
MTB > #residuals.
MTB > BRIEF=2
MTB > REGRESS Y values in C2 on 1 predictor in C1;
SUBC> RESIDUALS into C3.

The regression equation is
Score(Y) = 42.4 - 1.59 Fsize(X)

Predictor        Coef      Stdev     t-ratio         p
Constant       42.449      1.021       41.57     0.000
Fsize(X)      -1.5874     0.1388      -11.44     0.000

s = 1.659     R-sq = 92.9%     R-sq(adj) = 92.2%

Analysis of Variance

SOURCE        DF          SS          MS          F         p
Regression     1      360.36      360.36     130.90     0.000
Error         10       27.53        2.75
Total         11      387.89
```

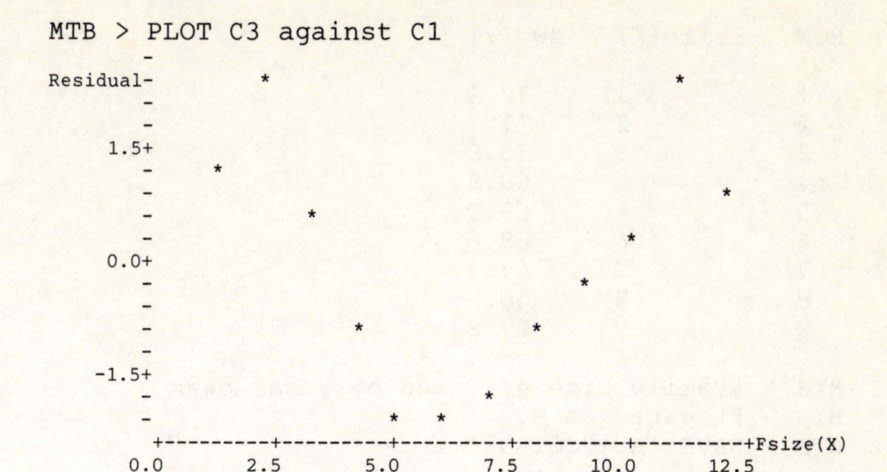

We note that there is an extremely large negative correlation of -0.96 between the two variables family size and mean test score. This leads to a large R^2 of 93%.

The equation test score (Y) = 42.4 − 1.59 family size (X) fits the data very well, the residuals being small and virtually patternless about zero.

This strong negative association between family size and the average score of children in each family grouping was one of the main findings of this very large study.

It is likely that both the variables average test score and family size are related to other socio-economic factors, in particular to the social class of the child.

(ii)

```
MTB > #Data from the 1947 Scottish Mental Survey.
MTB > #Mean weights of 3404 boys grouped by family
MTB > #size. Enter the data in the Editor and name
MTB > #the columns C1 Fsize(X) and C2 Mwt(Y)
MTB > PRINT C1 C2
```

```
ROW   Fsize(X)   Mwt(Y)

  1          1     72.1
  2          2     71.7
  3          3     70.9
  4          4     69.5
  5          5     69.3
  6          6     68.8
  7          7     68.7
  8          8     68.5
  9          9     67.8

MTB > #Family size of 9 and over has been
MTB > #treated as 9.
MTB > SAVE 'A:\SCOTS02'

Worksheet saved into file: A:\SCOTS02.MTW
MTB > #Above saves data.
MTB > PLOT C2 against C1
```

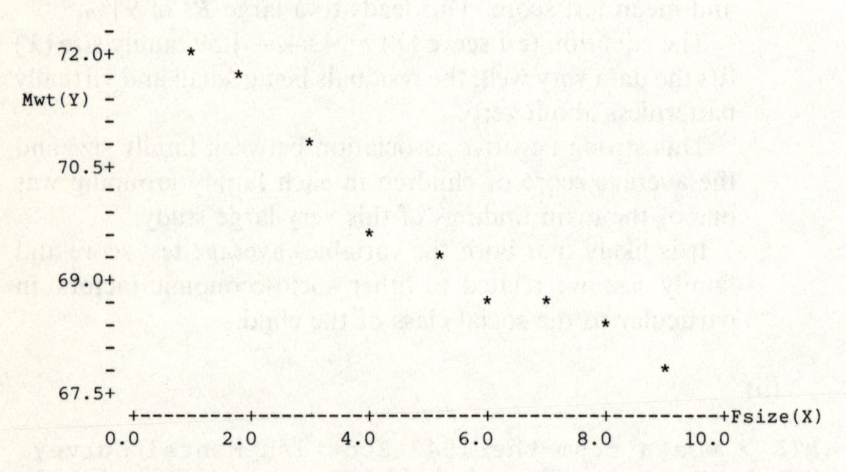


```
MTB > CORRELATE C2 and C1

Correlation of Mwt(Y) and Fsize(X) = -0.964

MTB > #Use least-square regression command
MTB > #and collect residuals.
MTB > BRIEF=2
MTB > REGRESS Y values in C2 on 1 predictor in C1;
SUBC> RESIDUALS into C3.
The regression equation is
Mwt(Y) = 72.4 - 0.532 Fsize(X)
```

```
Predictor        Coef       Stdev    t-ratio        p
Constant      72.3583      0.3131     231.14    0.000
Fsize(X)     -0.53167     0.05563      -9.56    0.000

s = 0.4309     R-sq = 92.9%     R-sq(adj) = 91.9%

Analysis of Variance

SOURCE         DF        SS         MS         F        p
Regression      1    16.960     16.960     91.34    0.000
Error           7     1.300      0.186
Total           8    18.260

MTB > #Go into editor and name C3 Residual.
MTB > PLOT C3 against C1
```

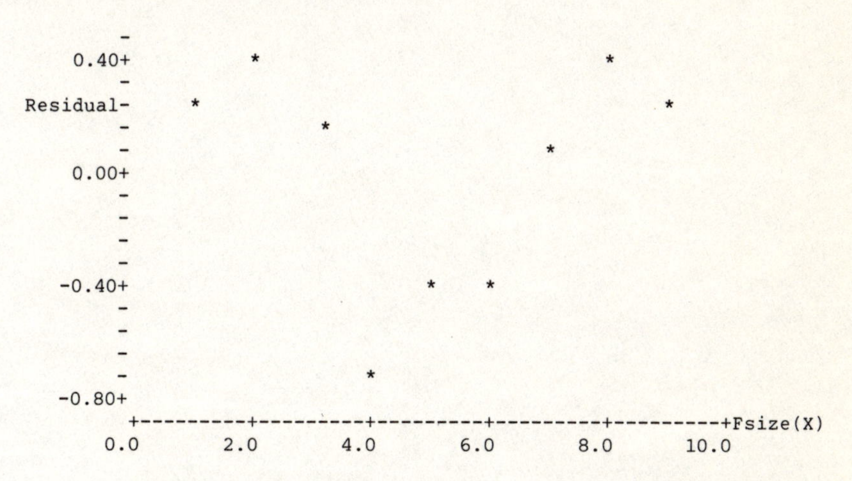

Again a very high negative correlation coefficient of -0.96 exists between the two variables family size and mean weight. This results in an R^2 value of 93%.

The equation mean weight (Y) = 72.4 − 0.532 family size (X) fits the data well. The residuals are small and without a distinct pattern.

As we observed in exercise 4 (i) family size and average intelligence scores are strongly negatively correlated, and the same is true of family size and the average weight of the representative sample of boys considered here.

(Indeed in the study a similar degree of negative correlation exists for both girls and boys in relation to both height and weight.) For the boys' data we can say that the least-squares regression line gives an extremely good fit to the data.

Again as we noted for family size and mean intelligence scores, it is likely that the variables of family size and average height that we are considering here are related to other socio-economic factors, and in particular to the social class of the child. In Britain height is fairly strongly correlated with social class but the relationship is not a linear one.

8 Linear Transformations

In this chapter we look at a way to apply the method of line fitting to data which has a curvilinear pattern. We learn how to 'unbend' the data and so greatly increase the scope of the powerful line fitting technique that we studied in chapter 7.

Consider for a moment the population growth shown in the first 100 years of Census returns in England and Wales, i.e. from 1801 to 1901. The growth can be illustrated by a simple graph as shown in figure 8.1.

There seems to be a distinct pattern to the data indicating a relationship between the two variables time X and population Y. If we wish to summarize this relationship in an equation we might be tempted to fit a straight line to the data; however, this would be unwise in view of the distinct curvilinear pattern visible in the plotted data. With this type of pattern in the raw data we could probably fit a quadratic equation such as $Y = aX^2 + b$ or even an exponential equation such as $Y = a2^x$. However, it will be simpler if we can unbend or straighten the data in some way so that we can apply the least-squares technique that we studied in chapter 7. A

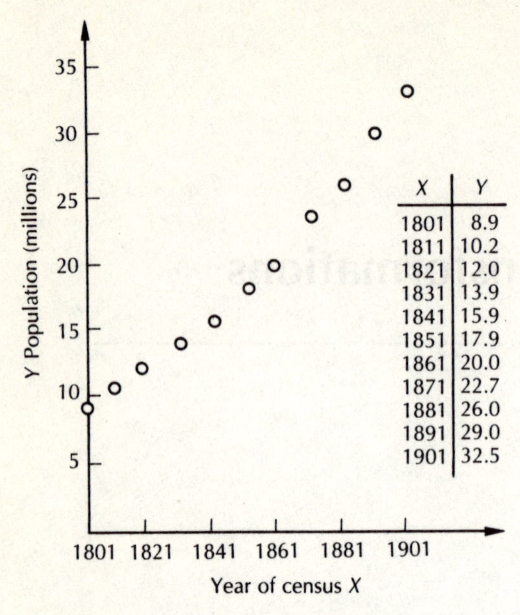

X	Y
1801	8.9
1811	10.2
1821	12.0
1831	13.9
1841	15.9
1851	17.9
1861	20.0
1871	22.7
1881	26.0
1891	29.0
1901	32.5

Source: *Population Trends* 69, 1992

Figure 8.1 Population growth

great deal of, but not all, curvilinear data can be transformed to a more linear form.

As we saw in the last chapter we can fit a straight line to summarize the data with a neat equation if the plotted data are relatively linear in form. If a scattergram of the data shows them to lie in a roughly lenticular or cigar-shaped pattern we can usually fit a simple straight line. Such a straight line links the two variables X and Y by a simple equation of the form $Y = aX + b$. By examining a scattergram of the data and the R^2 value of the least-squares analysis we can get a good idea of the goodness of fit of the line we have calculated. The R^2 value will tell us how much of the variability in the bivariate data is accounted for by the fitted line. When the fit is good R^2 square is high and most residuals to the fit are small.

However, if we have data that exhibit a curvilinear pattern when plotted we should not try to fit a straight line. As stressed previously, we must always examine the form of the data before using the method of least squares. Some curvilinear patterns are easily identified in the raw data but others are less easy to detect. If we do inadvertently fit a straight line to curvilinear data the residuals from the fit will show a curvilinear form. Indeed it is often easier to detect a curvilinear pattern in the residuals to the fit rather than in the scattergram of the raw data. As we stressed in the last chapter residuals should always be plotted and examined. It is usually best to plot them against the explanatory variable X, or if X has been transformed then against the transformed values of X.

Three Types of Transformation

Once we have detected the type or shape of the curvilinearity in the data we can apply various simple transformations to straighten them. There is a ladder or scale of such transforms which can be applied not only to different types of curvature but also to different degrees of curvature. However, here I focus attention, for simplicity, on three of the most common transformations.

When we examine the effect of these three transformations on data it is easier to see the effect if we deal with simple data sets and I will initially illustrate the method using small data sets. If the data have a curvilinear pattern in one of the forms in figure 8.2 we can usually successfully apply a transformation to one or both of the variables X and Y.

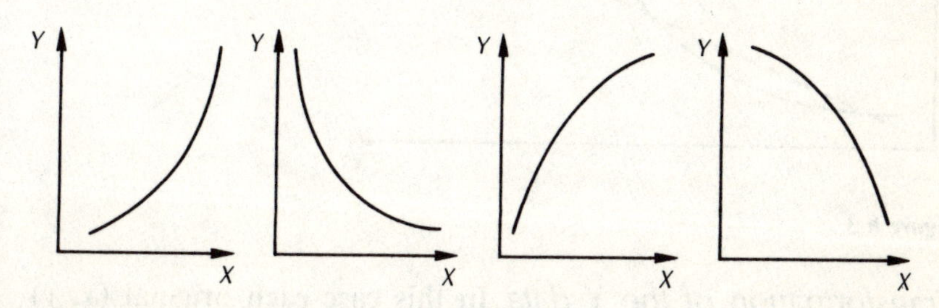

Figure 8.2

When we transform data what we are trying to do is to get more symmetry into the X variable, the Y variable or both. That is, we use a selected transform to 'pull' the data about and try to get it to fit a straight line. In general terms we can do three things with our selected transformation: transform the X data, transform the Y data or transform them both. Let us consider each of these three in turn.

Transformation of the X data Here each original (x, y) point of the bivariate data set is transformed or mapped to (x_{new}, y).

$$(x, y) \rightarrow (x_{new}, y)$$

This could be done, say, with the transformation x^2, where every x value in the data set is squared, i.e. multiplied by itself. In this case we would have the transform or mapping

$$(x, y) \rightarrow (x_{new}, y) = (x^2, y)$$

Each element of our original data set for the variable X has been changed from x to x^2. This is illustrated in figure 8.3.

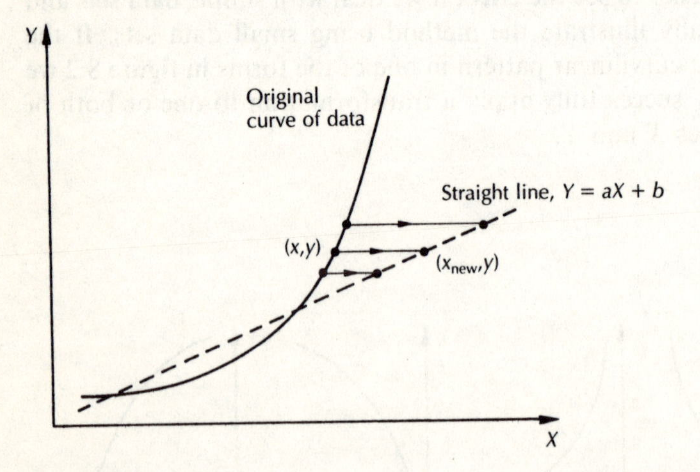

Figure 8.3

Transformation of the Y data In this case each original (x, y) point of the bivariate data set is transformed or mapped to (x, y_{new}).

$$(x, y) \rightarrow (x, y_{\text{new}})$$

This could be done for instance with the transformation $\sqrt{y}$ and we would have

$$(x, y) \rightarrow (x, y_{\text{new}}) = (x, \sqrt{y})$$

Each element of our original data set for the variable Y has been changed from y to $\sqrt{y}$. This is illustrated in figure 8.4.

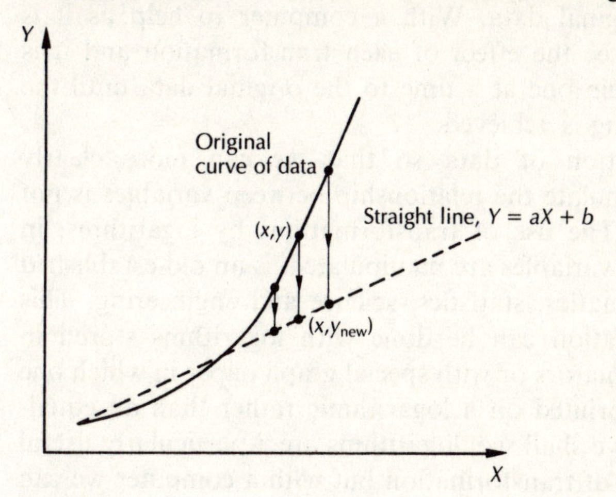

Figure 8.4

Transformation of both the X and the Y data Here each original (x, y) point in the bivariate data set is transformed or mapped to $(x_{new}, y_{\text{new}})$,

$$(x, y) \rightarrow (x_{\text{new}}, y_{\text{new}})$$

If this was done using the square root of the X variable and say the logarithm of the Y variable we would have

$$(x, y) \rightarrow (x_{\text{new}}, y_{\text{new}}) = (\sqrt{x}, \log y)$$

What we are doing in each of the above cases is replacing one or both variables by a function of that variable. Consequently we must also modify the straight-line equation. For instance, if we elect to transform only the X variable and replace X by the function X^2, i.e. each x value in the data is mapped or transformed to x^2, our straight-line equation $Y = aX + b$ will now be replaced by

$$Y = a(X^2) + b.$$

How much manipulation or transformation of the data is necessary will obviously depend on the degree of curvature or pattern in the original data. With a computer to help us it is relatively easy to see the effect of each transformation and it is worth applying them one at a time to the original data until the desired straightening is achieved.

The transformation of data so that we can more clearly recognize and formulate the relationship between variables is not a new technique. The use of transformations by logarithms, in which one or both variables are manipulated, is an old established practice in mathematics, statistics, science and engineering. This type of transformation can be done with logarithms stored in tables, pocket calculators or with special graph paper in which one or both axes are printed on a logarithmic rather than an equal-interval scale. As we shall see, logarithms are a particularly useful and common type of transformation but with a computer we can use a whole battery or scale of transformations with relative ease.

A Scale of Transformations

Let us now look at such a scale of transformations that we can apply to data in order to straighten them. John Tukey (1977) was an early pioneer in data analysis and he has collected a set of transformations together in a scale of strength as follows:

X variable starting point (at x)

$$\ldots -1/x \leftarrow -1/\sqrt{x} \leftarrow \log x \leftarrow \sqrt{x} \leftarrow x \rightarrow x^2 \rightarrow x^3 \rightarrow x^4 \ldots$$

Y variable starting point (at y)

$$\ldots -1/y \leftarrow -1/\sqrt{y} \leftarrow \log y \leftarrow \sqrt{y} \leftarrow y \rightarrow y^2 \rightarrow y^3 \rightarrow y^4 \ldots$$

The scale starts in the 'middle' with the original value of the variable and transformations become stronger in their effect the further we move away from the starting point either to the left or the right.

vstrong ← strong ← moderate ← mild ← original variable → mild → moderate → strong → vstrong
(x or y)

In using this scale of transformations we can imagine ourselves to be starting in the 'middle' (starting point) with the original data and then we can move outwards to the left or right trying various transformations of increasing strength until we find one that suits the task. If the mild transformations under-correct the data then we can try the moderate or maybe even the stronger ones. First try the Khorma, then the Madras and then maybe the Vindaloo.

In this introductory text I have restricted myself to three of the above transformations, these being the square root, the logarithm to the base 10 and the negative reciprocal of the square root. Each of these three transformations can of course be applied to each variable or both. The transformations when applied to the X and Y variables give us

$$x \rightarrow \sqrt{x} \qquad y \rightarrow \sqrt{y}$$
$$x \rightarrow \log x \qquad y \rightarrow \log y$$
$$x \rightarrow -1/\sqrt{x} \qquad y \rightarrow -1/\sqrt{y}$$

These are three of the most useful and popular transformations and they enable us to straighten out a great deal of data. As with any transformation we cannot always expect to be successful but in a large number of cases we can achieve a fair degree of linearity using these transformations with one or both variables. After applying one or more of the transformations to the raw data we can then plot the transformed data and if it appears to have a reasonable degree of linearity we can use our least-squares regression method to fit a straight line to it.

Before looking at the transformation of some simple data sets we look briefly at the general effect of the three transformations that I have chosen on just one side of a bivariate distribution.

As mentioned earlier the idea of transformation is to get a higher degree of symmetry into one or both of the data sets that make up a bivariate distribution. Let us imagine that one of the variables, X, in a bivariate distribution has a skewed or lopsided distribution. To make the data more symmetrical, i.e. to ensure that the mean and the median are reasonably close and the shape of the distribution on each side of these averages is more or less the same, we can imagine pulling the data over towards the thin tail with a suitable distribution (figure 8.5).

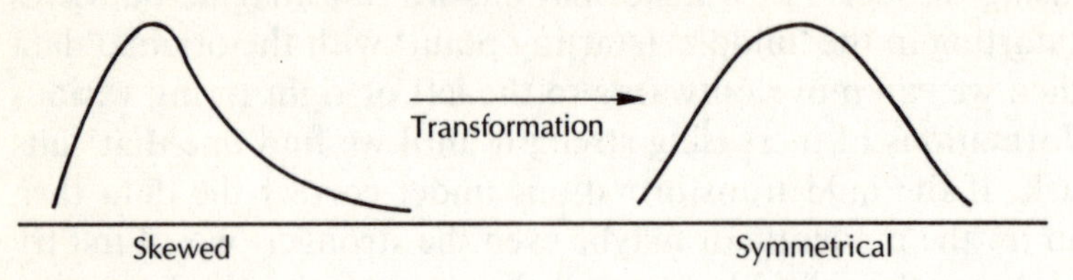

Figure 8.5

To illustrate the process let us simulate a very skewed distribution and then apply each of the chosen transformations to it in turn and note the effect. Imagine that we are dealing with half of the data in a bivariate distribution. We shall look at the distribution of a variable, say X, which has a very skewed distribution and see what effect our three transformations have on its shape.

In order to get a suitably lopsided distribution to manipulate we will simulate a χ^2 (Chi-square) distribution with one degree of freedom. We will look in detail at this type of distribution in chapter 11 but here it is sufficient to note that the χ^2 distribution, which is a close relative of the normal distribution (chapters 9 and 10) has a very skewed distribution when its degree of freedom is a small number, say 1, 2, 3 or 4. Don't worry at this stage what the term degree of freedom means – just note that it is a characteristic or parameter that determines the shape of the distribution. The χ^2 distribution is really a whole family of distributions and each member of the family has a characteristic shape which is determined by the number of degrees of freedom that it possesses. At this stage we are not interested in χ^2 as a type of distribution; we merely need a skewed distribution on which we can test our tools of transformation.

A Comparison of Transformations

As with previous examples, you should work this one through for yourself on the computer. If you want to save it don't forget to use a suitable OUTFILE command such as MTB > OUTFILE 'A:\TRANSF02'.

I have first generated a skewed distribution of the variable X in C1 and then transformed it by each of the transformations in turn, $\sqrt{x}$, $\log x$, $-1/\sqrt{x}$. Some comments have been embedded in the print-out and I give a brief overview at the end.

```
MTB > #This simulates a chi-square distribution
MTB > #with one degree of freedom V=1.
MTB > RANDOM 100 observations into C1;
SUBC> CHISQUARE V=1.
MTB > HISTOGRAM C1

Histogram of C1 N = 100 Each * represents 2 obs.

Midpoint   Count
       0      52 **************************
       1      23 ************
       2      10 *****
       3       4 **
       4       7 ****
       5       3 **
       6       0
       7       0
       8       0
       9       0
      10       1 *

MTB > #Note the very strong skew of the distribution.
MTB > MEAN C1
MEAN = 1.1701
MTB > MEDIAN C1
   MEDIAN = 0.48085

MTB > #First transformation, taking the square root,
MTB > #x→√x
MTB > LET C2 = SQRT (C1)
MTB > HISTOGRAM C2
```

```
Histogram of C2 N = 100

Midpoint  Count
     0.0     11 ***********
     0.4     33 *********************************
     0.8     20 ********************
     1.2     17 *****************
     1.6      7 *******
     2.0      8 ********
     2.4      3 ***
     2.8      0
     3.2      1 *

MTB > MEAN C2
   MEAN = 0.87158
MTB > MEDIAN C2
   MEDIAN = 0.69343
MTB > #Second transformation, taking logarithms
MTB > #to the base ten, x→log x
MTB > LET C3 = LOGT (C1)
MTB > HISTOGRAM C3

Histogram of C3 N = 100

Midpoint  Count
     -9      1 *
     -8      0
     -7      0
     -6      0
     -5      1 *
     -4      1 *
     -3      0
     -2      8 ********
     -1     30 ******************************
      0     47 ***********************************************
      1     12 ************

MTB > MEAN C3
   MEAN = -0.51540
MTB > MEDIAN C3
   MEDIAN = -0.31800
MTB > #This second transformation is an over
MTB > #correction but I will make the third
MTB > #transformation so that you can see the
MTB > #effect.
MTB > #This is the third transformation, x→-1/x
MTB > LET C4 = -1/SQRT (C1)
```

```
MTB > HISTOGRAM C4

Histogram of C4 N = 100
Each * represents 2 obs.

Midpoint Count
  -24000     1 *
  -22000     0
  -20000     0
  -18000     0
  -16000     0
  -14000     0
  -12000     0
  -10000     0
   -8000     0
   -6000     0
   -4000     0
   -2000     0
       0    99 **************************************************

MTB > MEAN C4
   MEAN = -251.20
MTB > MEDIAN C4
   MEDIAN = -1.4421
MTB > #Very radical effect. Strong skew in the
MTB > #opposite direction. Very much over
MTB > #corrected.
```

The χ^2 distribution that we have simulated has a very skewed distribution as you can see by the first histogram. There is a wide separation of the two measures of centre or average: the mean is 1.2 whilst the median at 0.48 is less than half of this value.

The first transformation, $x \rightarrow \sqrt{x}$, has a considerable effect on the distribution: it becomes less skewed and the mean and the median move closer together (mean = 0.87, median = 0.69).

The second transformation, $x \rightarrow \log x$, has over-corrected the original skew and introduced a skew in the opposite direction.

As the third transformation, $x \rightarrow -1/\sqrt{x}$, is stronger in its effect on the distribution than the second, which has over-corrected it, we do not need to continue along the scale of transformations. We have already gone too far. However, I have done so in order that you can see the effect of the third of our chosen transformations. The distribution is now very radically skewed in the opposite direction to the original.

Of the three transformations, the first, $x \rightarrow \sqrt{x}$, seems to be the most successful in reducing the asymmetry of the distribution, but this is also a slight under-correction.

The above exercise clearly shows the effect of our three transformations on *one* skewed distribution; however, we have to deal with bivariate data, i.e. X and Y values. Now that we have an idea of how the transformations work let us try them on some simple bivariate data. We look at some of the effects of the three chosen transformations on the independent variable X, on the dependent variable Y, and on them both.

In order to simplify matters so that we can see more clearly what is going on I have chosen a small subset of the suicide data in DATA04.DAT. I have taken the male suicide rates for Austria and plotted them against the age midmarks. The distribution fragment is as follows:

Age (X) in years	20	30	40	50	60	70	80
Suicide rate (Y)	29.3	42.9	44.8	54.9	48.5	68.1	125.2

Please work the exercise through for yourself and note the comments embedded in the print-out. If you wish to save your work don't forget to use an OUTFILE command such as MTB > OUTFILE 'A:\TRANSF03' at the start. A brief overview of the exercise is given later.

```
MTB > #Type the data into the Editor and
MTB > #name the columns C1 X C2 Y
MTB > #These are the 4 transformations.
MTB > LET C3=SQRT (C1) #First transform        x→√x
MTB > LET C4=SQRT (C2) #Second transform       y→√y
MTB > LET C5=LOGT (C1) #Third transform      x→log x
MTB > LET C6=LOGT (C2) #Fourth transform     y→log y
MTB > PRINT C1-C6

ROW    X      Y   SQRT(X)    SQRT(Y)   LOGT(X)   LOGT(Y)

  1   20   29.3   4.47214     5.4129   1.30103   1.46687
  2   30   42.9   5.47723     6.5498   1.47712   1.63246
  3   40   44.8   6.32456     6.6933   1.60206   1.65128
  4   50   54.9   7.07107     7.4095   1.69897   1.73957
  5   60   48.5   7.74597     6.9642   1.77815   1.68574
  6   70   68.1   8.36660     8.2523   1.84510   1.83315
  7   80  125.2   8.94427    11.1893   1.90309   2.09760
```

```
MTB > #6 plots follow
MTB > PLOT C2 against C1
```

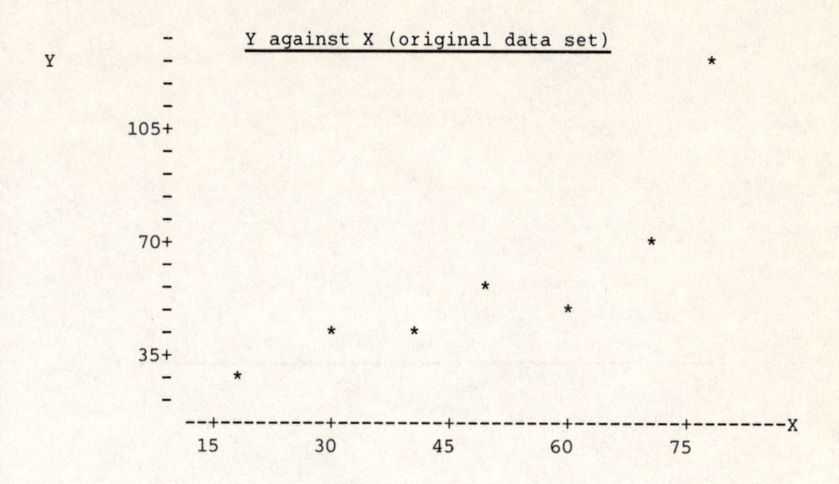


```
MTB > PLOT C2 against C3
```

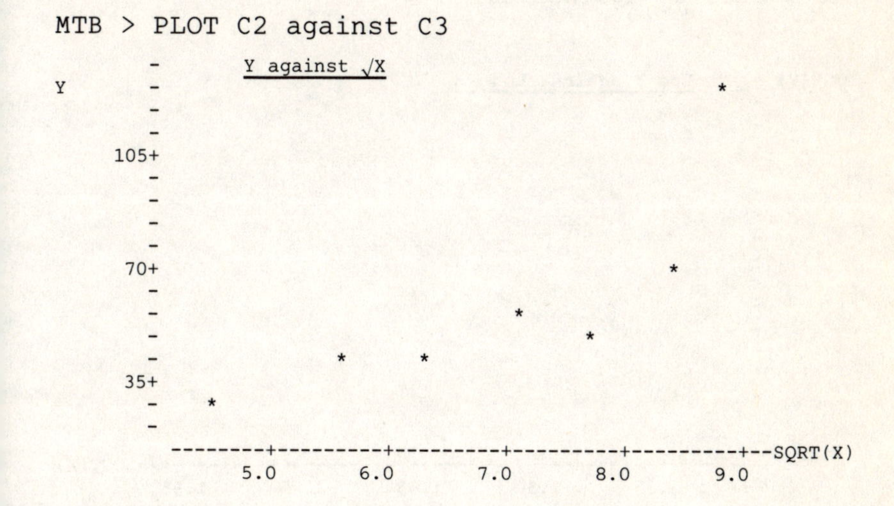

MTB > PLOT C4 against C1

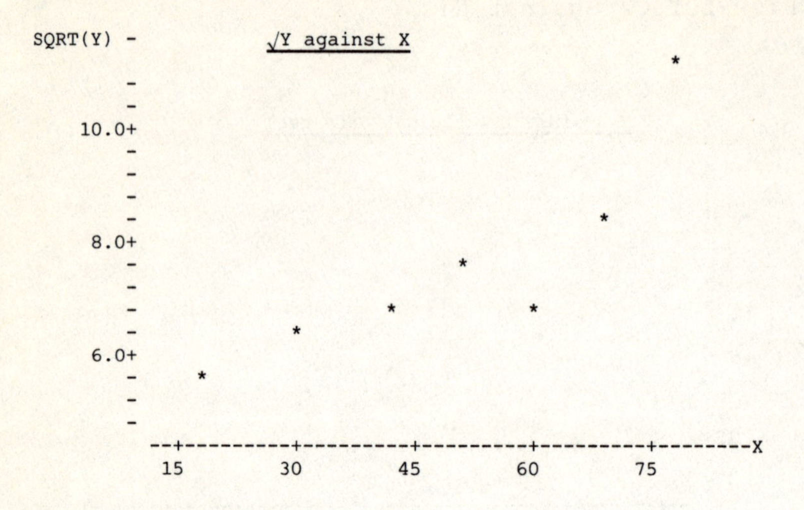

MTB > PLOT C6 against C5

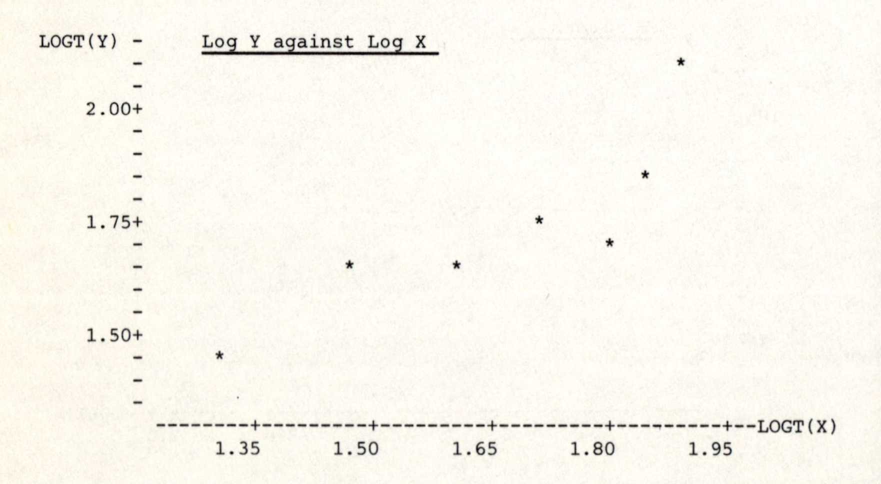

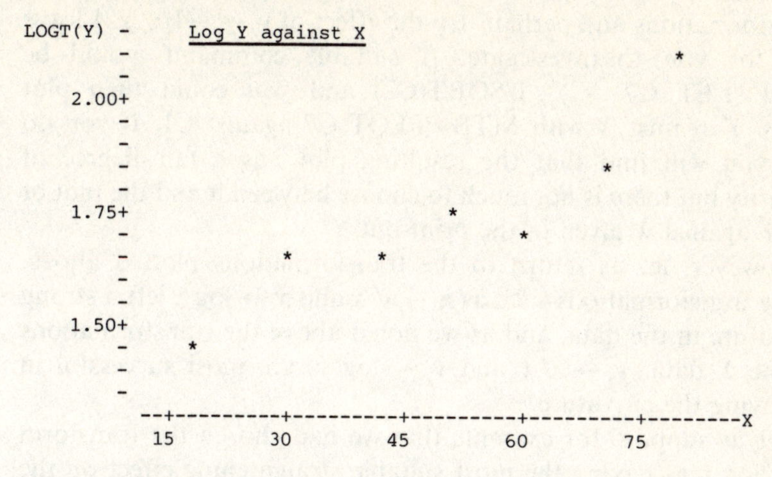


```
MTB > PLOT C2 against C5
```

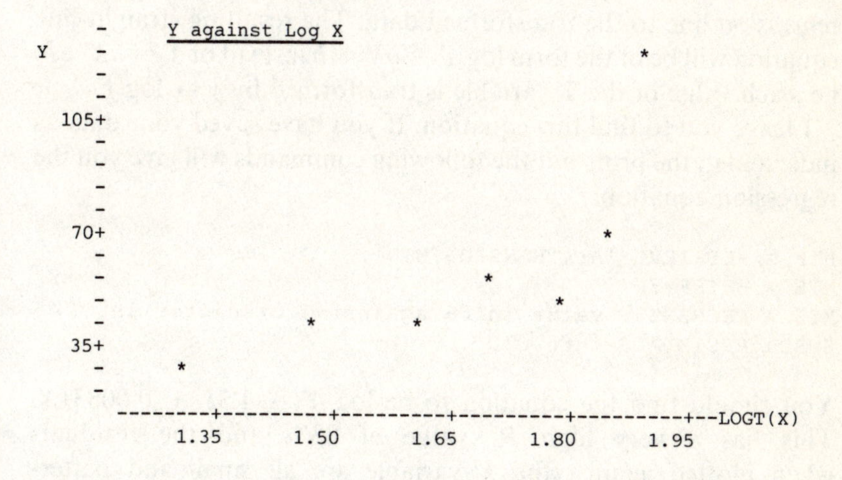

As you can see, the original data have a very strong J form, curving upwards from lower left to upper right of the graph. To achieve some linearity we need to pull the upper end of the J curve down towards the *X* axis or away from the *Y* axis.

In the five plots chosen for illustration you will see that two seem to have achieved a reasonable degree of straightening, i.e. $\sqrt{Y}$ against *X*, and log *Y* against *X*. That is, the transformation of the *Y* variable seems to give the best results here. If you look back at the scales of transformation (page 216) this suggests that we could try to move further to the left along the Y scale of transformations and perhaps try the effect of $y \rightarrow -1/\sqrt{y}$. I leave this for you to investigate. A suitable command would be MTB>LET C7 = −1/SQRT(C2) and you could then plot $-1/\sqrt{Y}$ against X with MTB>PLOT C7 against C1. If you do this you will find that the resulting plot has a fair degree of linearity but there is not much to choose between it and the plot of log Y against X given in the print-out.

However, let us return to the transformations plotted above. Some transformations such as $x \rightarrow \sqrt{x}$ and $x \rightarrow \log x$ left a strong curvature in the data, and as we noted above the transformations of the Y data, $y \rightarrow \sqrt{y}$ and $y \rightarrow \log y$, are most successful in removing the curvature.

Let us suppose for example that we had chosen the transform $y \rightarrow \log y$ as having the most suitable straightening effect on the original data and that we now wish to proceed to fit a least-squares regression line to the transformed data. The resulting straight-line equation will be of the form $\log Y = aX + b$ instead of $Y = aX + b$, i.e. each value of the Y variable is transformed by $y \rightarrow \log y$.

I leave you to find this equation. If you have saved your data as indicated in the print-out the following commands will give you the regression equation:

```
MTB > RETRIEVE 'A:\TRANSF03.MTW'
MTB > BRIEF=3
MTB > REGRESS Y value in C6 against 1 predictor in C1;
SUBC> RESIDUALS into C10.
```

You should find the equation to be $\log Y = 1.31 + 0.00831X$. This has a very high R^2 value of 83% and the residuals when plotted against the X variable are all small and patternless. For males in Austria at this time the two variables of age and suicide rate can be related by the equation logarithm (suicide rate) = 1.31 + 0.00831 × age.

Let us now look again at the scales of transformation on page 216. With the J curve that we have just considered we found that some of the transformations helped us achieve linearity in the data but others did not. We have a problem in knowing which type of transformation to use on various types of curve and in what strength. Where on the scale of transformations should we start? Should we transform X or Y, or both?

As mentioned previously it is generally best to start in the middle of the scales and work our way outwards to the left or right and thereby increase the strength or effect of the transformation. However, different types of curvature in the original data require different treatment.

Type of curve		Possible X transforms	Possible Y transforms
Type 1		$x \rightarrow \sqrt{x}$ or $\log x$ etc.	$y \rightarrow y^2$ or y^3 etc.
Type 2		$x \rightarrow \sqrt{x}$ or $\log x$ etc.	$y \rightarrow \sqrt{y}$ or $\log y$ etc.
Type 3		$x \rightarrow x^2$ or x^3 etc.	$y \rightarrow \sqrt{y}$ or $\log y$ etc.
Type 4		$x \rightarrow x^2$ or x^3 etc.	$y \rightarrow y^2$ or y^3 etc.

Figure 8.6

If we consider data with four types of curvature as indicated in figure 8.6 we can apply transformations to the X data and Y data as shown; we can then, should the need arise, also apply the same transformations to *both* the X and Y data sets. In each case we should start in the middle of the scale of X or Y transformations and work outwards as indicated.

The Austrian male suicide data that we looked at had a curvature of type 3 and we were able to straighten it successfully using a transformation of the Y data. However, as you can see from the above table, we could also have tried transforming the X variable with x^2 or x^3 etc. The use of the abbreviation etc. in the above table indicates that if the first two transformations do not have the desired straightening effect then we should try those further out in the direction indicated, i.e. stronger transformations further along the scales.

Let us now turn our attention to a curvature of type 2 and see if we can straighten it. I have again used a small data set fragment so that we can see what is going on.

The data concern the incidence of low birthweight and the height of the mother in Scotland in 1979 and are part of a very large and comprehensive study by Alison Macfarlane and Miranda Mugford (1984, p. 106).

Maternal height X (cm)	146	153	158	163	168	172	180
Percentage birthweights less than 2500 g (Y)	16	10.0	8.3	5.8	4.0	3.3	1.9

As you probably know, low birthweight is strongly negatively associated with high perinatal and infant mortality, and as can be seen from the data the percentage of underweight babies increases as the mothers' height falls. When plotted the distribution of data will be of type 2. Although maternal height is strongly related to social class in Britain (the higher the class, the taller the mother) the relationship shown in this table between mothers' height and percentage of low birthweight children, and so higher risk to infants, holds good across all social classes.

Let us plot and find a suitable transformation for the data and then find a linear equation to describe the relationship between the two variables. As previously I would like you to work through the example for yourself on the computer. If you wish to record your analysis do not forget to OUTFILE your work with a suitable command at the beginning, for instance
MTB > OUTFILE 'A:\TRANSF06'.

As the data have a type 2 curvature, by reference to our scale of transformations we can start with either $x \rightarrow \surd$ or $y \rightarrow \sqrt{y}$ and work outwards. I have only employed four of these transformations in my search, these being the first two for the X variable and the first two for the Y variable. However, for brevity I will not reproduce all the possible plots from these four transformations but focus instead on just two plots: (i) X against log Y and (ii) log X against LOG Y. Both of these seem to straighten the data quite well, but I have used the latter to calculate a regression line.

```
MTB > #Type the data into the editor or
MTB > #read it in from the data file.
MTB > READ 'A:\DATA06.DAT' into C1-C2
MTB > #The transforms are as follows
MTB > LET C3=SQRT (C1) # x→x
MTB > LET C4=SQRT (C2) # y→√y
MTB > LET C5=LOGT (C1) # x→log x
MTB > LET C6=LOGT (C2) # y→log y
MTB > #Go into the editor and name all the columns.
MTB > PRINT C1-C6
```

ROW	X	Y	SQRT(X)	SQRT(Y)	LOG(X)	LOG(Y)
1	146	16.0	12.0830	4.00000	2.16435	1.20412
2	153	10.0	12.3693	3.16228	2.18469	1.00000
3	158	8.3	12.5698	2.88097	2.19866	0.91908
4	163	5.8	12.7671	2.40832	2.21219	0.76343
5	168	4.0	12.9615	2.00000	2.22531	0.60206
6	172	3.3	13.1149	1.81659	2.23553	0.51851
7	180	1.9	13.4164	1.37840	2.25527	0.27875

```
MTB > SAVE 'A:\TRANSF06'
MTB > #This saves the new transformed data.
Worksheet saved into file: A:\TRANSF06.MTW
MTB > PLOT C2 against C1
```

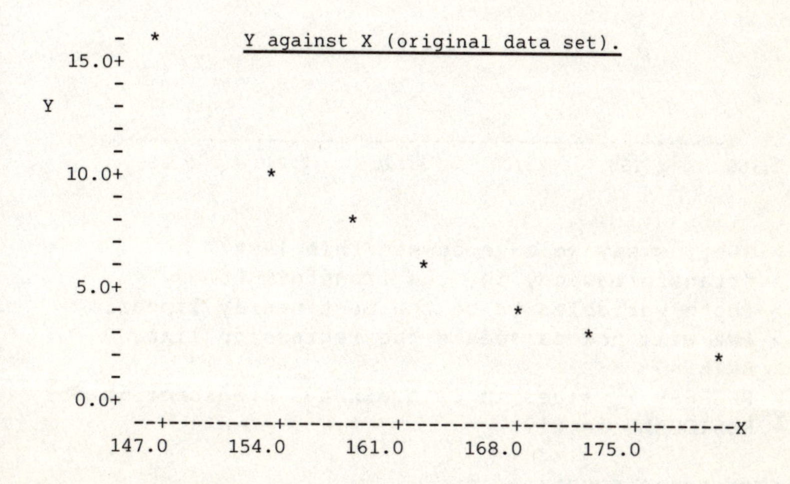

MTB > PLOT C6 against C1

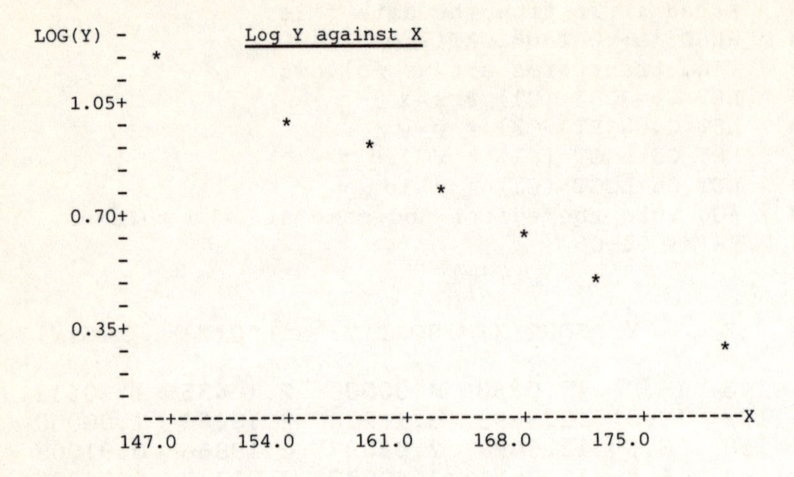

MTB > PLOT C6 against C5

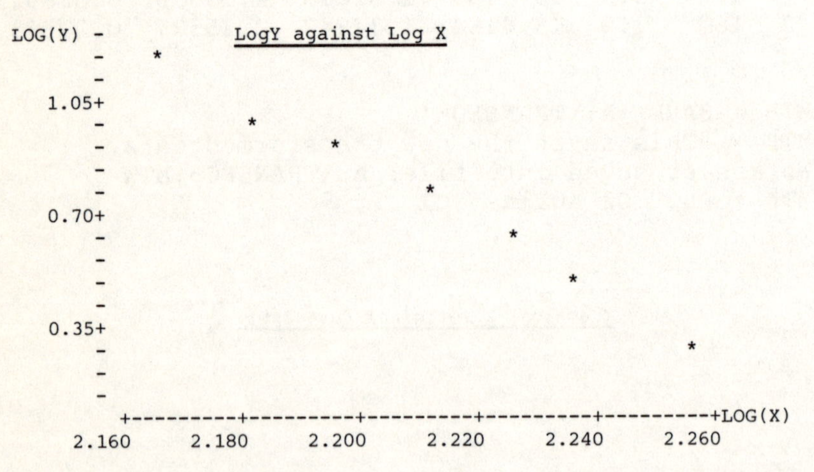


```
MTB > #Let us say we have chosen this last
MTB > #transformation, ie. the transform of
MTB > #both variables to be the most nearly linear.
MTB > #We will now calculate the regression line.
MTB > BRIEF=3
MTB > REGRESS Y values in C6 against 1 predictor in C5;
SUBC> RESIDUALS in C10.
```

The regression equation is
LOG(Y) = 23.1 - 10.1 LOG(X)

```
Predictor        Coef       Stdev     t-ratio        p
Constant      23.0940      0.8134       28.39    0.000
LOG(X)       -10.1042      0.3679      -27.47    0.000

s = 0.02799     R-sq = 99.3%      R-sq(adj) = 99.2%

Analysis of Variance

SOURCE        DF        SS         MS         F       p
Regression     1   0.59094    0.59094    754.46   0.000
Error          5   0.00392    0.00078
Total          6   0.59485

Obs.  LOG(X)  LOG(Y)     Fit  Stdev.Fit  Residual  St.Resid
  1     2.16  1.2041  1.2250     0.0201   -0.0209     -1.07
  2     2.18  1.0000  1.0195     0.0143   -0.0195     -0.81
  3     2.20  0.9191  0.8784     0.0115    0.0407      1.59
  4     2.21  0.7634  0.7417     0.0106    0.0217      0.84
  5     2.23  0.6021  0.6091     0.0118   -0.0070     -0.28
  6     2.24  0.5185  0.5059     0.0139    0.0127      0.52
  7     2.26  0.2788  0.3064     0.0195   -0.0276     -1.37
```

The earlier transformations in the scales $x \to \sqrt{x}$, $y \to \sqrt{y}$ and $x \to \log x$, when used with the other variable, did not give a satisfactory straightening of the data. However, both of the transformations illustrated in the print-out, log Y against X and log Y against log X, give good results. There is not much to choose between them, but I have chosen the double logarithmic transformation to regress.

The resulting regression equation is $\log Y = 23.1 - 10.1 \log X$ and it has a very high R^2 value of 99% and small residuals without a pattern. An equation of the relationship between the two variables is therefore

Log (% low b.weight) = 23.1 − 10.1 log (maternal height)
(logarithms to the base 10 and heights in centimetres).

Let us end this chapter by looking at a much larger data set, DATA11.DAT, and try to formulate an expression for the relationship between the variables of wealth, as measured by GNP per capita, and health, as measured by the index of infant mortality.

As usual I would like you to work this through for yourself using the print-out as a guide. I have embedded some notes in the print-out. With four transformations of each variable plus the original variables we can get 25 plots (24 transform plots plus one original). You will notice that, for brevity, I have only shown three of the transformed plots. I have indicated the other possible plots in an orderly fashion, however, and you may find it instructive to work your way through them if your computer is reasonably large/fast. Most of these transformations over- or under-correct the data and if you plot them in the order I have indicated you can observe the progressive changes in the form of the plotted data as we work outwards on the two scales of transformation.

I give a further commentary on the transformations and regression line analysis after the print-out. It is worth saving your analysis with a suitable OUTFILE command such as MTB > OUTFILE 'A:\TRANSF07', and after you have made the transformations you will probably wish to save the transformed data sheet with a SAVE command such as MTB > SAVE 'A:\TRANSF07'.

```
MTB > #Read file DATA11.DAT into the editor.
MTB > READ 'A:\DATA11.DAT' into C1-C12
     59 ROWS READ

ROW  C1    C2  C3  C4  C5  C6    C7  C8  C9  C10

  1   1    80   5  10  35  17   173  45  21  300
  2   1   610   6  15  34   *   173  56  29    *
  3   1   180   7  11   7  17   144   *   *  170
  4   1   120   *   *   *  34   130   *   *    *
 . . .

ROW  C11   C12

  1   48  15.7
  2   46  10.0
  3   48   8.8
  4   46  49.2
 . . .

MTB > #Go into editor and name C2 GNP(X)
MTB > #and C7 IMR(Y)
MTB > PLOT C7 against C2
```

```
     -           Y against X (original data)
 180+
     - **
  C7 -
IMR(Y)- *
     - *
 120+ 2
     - *3
     - 2*
     -  **
     -  4* 2     *
  60+   2  2      *
     - ***   *
     -  **2*
     -  ****                       *
     -   2 *        *2   *   2*    *   2**          *
   0+                                         *
      +---------+---------+---------+---------+---------+C2 GNP(X)
      0       6000     12000     18000     24000     30000
```

```
MTB > #It is a Type 2 curve with a very strong
MTB > #curvature.
MTB > #First make the X transforms with the LET
MTB > #commands.
MTB > LET C13= SQRT (C2)      # x→√x
MTB > LET C14= LOGT (C2)      # x→log x
MTB > LET C15= -1/SQRT (C2) # x→-1/√x
MTB > LET C16= -1/(C2)        # x→-1/x
MTB > #Now make the Y transforms.
MTB > LET C17= SQRT (C7)      # y→√y
MTB > LET C18= LOGT (C7)      # y→log y
MTB > LET C19= -1/SQRT (C7) # y→-1/√y
MTB > LET C20= -1/(C2)        # y→-1/√y

MTB > #Go into editor and name the columns.
MTB > #Save the transformed data.
MTB > SAVE 'A:\TRANSF07'

Worksheet saved into file: A:\TRANSF07.MTW
MTB > #The possible plots using the above
MTB > #transforms are as indicated below. You
MTB > #should try as many as you can.
MTB > #PLOT in order given (work down the columns)
MTB > #to see progressive shifts of variables.

MTB > #            C17 & C2    C18 & C2    C19 & C2    C20 & C2
MTB > # C7 & C13   C17 & C13   C18 & C13   C19 & C13   C20 & C13
MTB > # C7 & C14   C17 & C14   C18 & C14   C19 & C14   C20 & C14
MTB > # C7 & C15   C17 & C15   C18 & C15   C19 & C15   C20 & C15
MTB > # C7 & C16   C17 & C16   C18 & C16   C19 & C16   C20 & C16
```

MTB > PLOT C7 C15

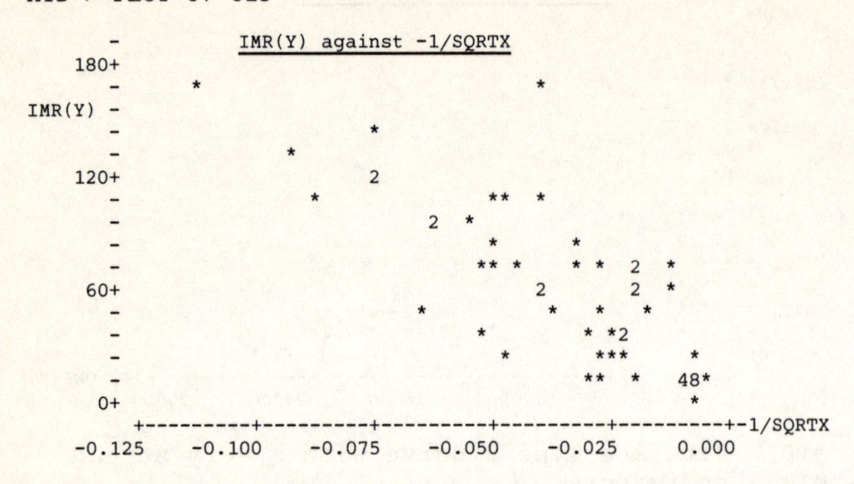

MTB > PLOT C17 C15

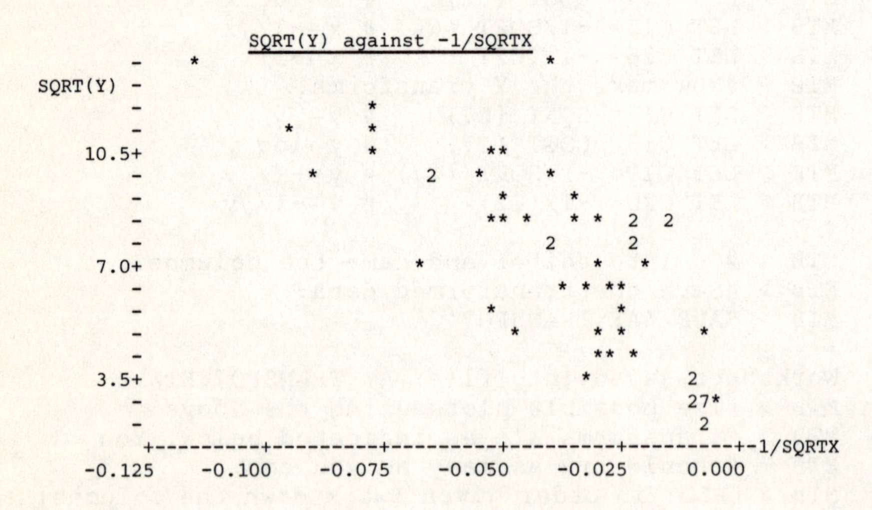

```
MTB > PLOT C18 C14
```

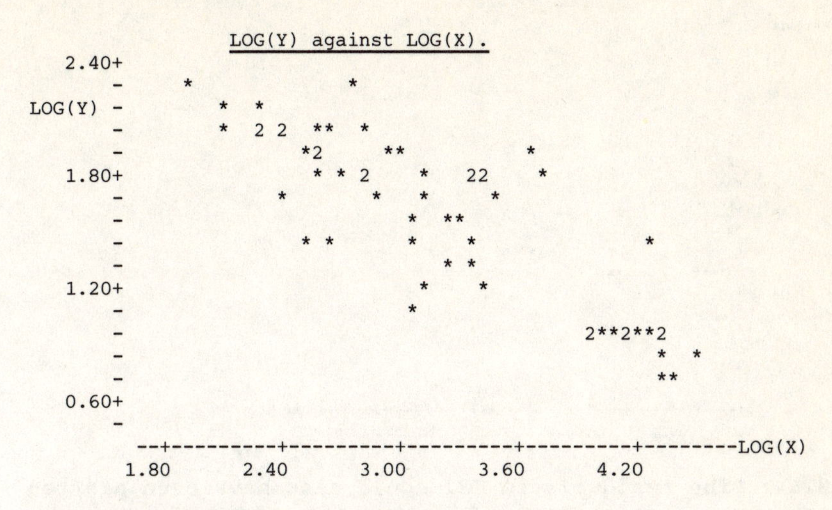


```
MTB > #We will now find the regression line
MTB > #using the transforms C18 LOG(Y) & C14 LOG(X).
MTB > BRIEF=2
MTB > REGRESS Y values in C18 against 1 predictor in C14;
SUBC> RESIDUALS into C21.
```

```
The regression equation is
LOG(Y) = 3.27 - 0.534 LOG(X)

Predictor      Coef      Stdev    t-ratio      p
Constant     3.2697     0.1477      22.14  0.000
LOG(X)     -0.53402    0.04492     -11.89  0.000

s = 0.2442    R-sq = 71.3%    R-sq(adj) = 70.8%

Analysis of Variance

SOURCE       DF       SS       MS        F      p
Regression    1   8.4295   8.4295   141.36  0.000
Error        57   3.3990   0.0596
Total        58  11.8285

Unusual Observations
Obs.  LOG(X)  LOG(Y)     Fit  Stdev.Fit  Residual  St.Resid
 17     3.73  1.8751  1.2804     0.0393    0.5947     2.47R
 22     3.78  1.8129  1.2513     0.0408    0.5616     2.33R
 45     3.07  1.0414  1.6312     0.0324   -0.5898    -2.44R

R denotes an obs. with a large st. resid.
```

```
MTB > PLOT C21 against C2

Residual-
       -         **
       - *
   0.40+    *                        *
       -    3
       - *2  *
       - 2
       - *3*2
   0.00+ 54*  *                          *              *
       - *2            *    *    *   *
       - *    *     *      2         * *
       -   ***      **                     *
       - *
  -0.40+  **
       - *
       -  *
       -
         +---------+---------+---------+---------+---------+GNP(X)
         0       6000     12000     18000     24000     30000

MTB > #The residuals in C21 could also have been plotted
MTB > #against the transformed X variable ie.
MTB > #LogX (C14).
```

From the plot of the original data we note that it has a very strong but curved pattern. It is a curve of type 2. The variables of wealth and infant mortality are clearly related but need to be transformed if we wish to describe the relationship by a straight line. When a country's wealth (GNP per capita) is low we note that infant mortality is high. For the rich industrialized nations the infant mortality is very low, usually below 10; however, for the poorest group of countries infant mortality is much greater. The lowest rate per 1000 live births is 5 in Japan while the highest is 173 in Mozambique and Angola.

The original data are a type 2 curve and I have chosen four appropriate transformations for each variable. However, as you can see I have only shown three of the resulting plots. These are the ones which seem to give the best results. Of the plotted transformations the one that uses Log X and Log Y has been used to calculate a regression line. I have listed the other possible plots of the transformed variables, however, and you should look at a few of them and observe the progressive change in the plotted data, usually from under-correction to over-correction.

The least-squares regression line for the chosen transformation has equation $\log Y = 3.27 - 0.534 \log X$ (logarithms to the base 10), i.e. $\log(\text{IMR}) = 3.27 - 0.534 \log(\text{GNP})$. The R^2 value is quite high at 71% and we note that most of the residuals are relatively small and grouped about the zero line without a distinctive pattern.

From what we have done we can conclude that the variables of wealth and infant mortality are intimately related. There are many intermediate social and economic factors, however, such as spending on health, the distribution of wealth, literacy (especially female literacy) and many significant urban/rural divisions, that have not been investigated.

Postscript

In this chapter we have outlined a very useful type of transformation for data. Using these techniques we can transform many curvilinear data and then apply the powerful methods of linear regression analysis to obtain a relatively simple equation linking the two variables.

In the next chapter we will change tack and look at the normal distribution and some of its close approximations.

Exercises

1 This example concerns data collected by F. L. Goodenough in a study of child development reported in *Motivation and Emotion*, (Young, 1966, p. 451). The study looked at the development of retaliative behaviour as a response to frustration in young children. Over a period, 2124 outbursts of anger were observed in a sample of 45 children of ages ranging between 7 months and nearly 8 years. The outbursts of anger were of two types, one in which the anger was general and undirected and the other in which anger was directed and retaliative.

Random and undirected anger covered such acts as jumping up and down, holding the breath, stamping and kicking, throwing oneself on the floor, screaming etc. Throwing objects, grabbing, pinching, biting, calling names, arguing, hitting etc. was characterized as directed retaliatory behaviour. It was found that acts of the first kind, undirected anger, became less frequent with advancing age while behaviour that was directed and retaliatory became more frequent. Normal development in children is marked by the increasing direction of anger against the specific frustrating object or person. The data are shown in table 8.1.

Table 8.1

Mean age in years (*X*)	*% Outbursts with undirected energy* (Y1)	*% Outbursts with directed energy* (Y2)
0.5	88.9	0.7
1.5	78.4	6.3
2.5	75.1	10.6
3.5	59.9	25.6
6.0	36.3	28.0

(i) Read the data into Minitab with the command MTB > READ 'A:\DATA08.DAT' into Cl-C3 and name the columns Cl Age(X), C2 Und(Y), C3 Dir(Y).

(ii) Plot both kinds of outburst against the independent variable X. Name each type of curve.

(iii) Indicate what types of transformation may be appropriate to unbend the data in each case.

(iv) Assume that in the case of the data for undirected anger the transformation $x \rightarrow x^2$ is suitable. Transform the data and calculate the least-squares regression line.

(v) Assume that in the case of the data for directed anger the transformation $x \rightarrow \log x$ (logarithms to the base 10) is suitable. Transform the data and calculate the least squares regression line.

2 The fuel consumption of the ship M.V. Baltic Sprite of London, within its practical operating range of speed, is shown in table 8.2

(i) Enter the data into the Minitab Editor and name the columns Cl Knots(X), C2 Tons(Y).

(ii) Plot the data and name the type of curve; indicate what kinds of transformations may be appropriate.

Table 8.2

Speed in knots (X)	*Consumption in tons (Y)*
10	12.2
11	14.9
12	18.1
13	24.2
14	33.8
15	42.5
16	51.7
17	59.9
18	70.1
18.5	79.7

(iii) Assume that the transformation $y \to \sqrt{y}$ is appropriate to straighten the data. Transform the data and calculate the least-squares regression line $Y = aX + b$.

3 The number of abortions in thousands, residents only, in England and Wales was as in table 8.3.

Table 8.3

Year (X)	*Thousands of women (Y)*
1971	94.6
1976	101.9
1981	128.6
1983	127.4
1984	136.4
1985	141.1
1986	147.6
1987	156.2
1988	168.3
1989	170.5
1990	173.9
1991	167.4

Source: *Population Trends*, 69, 1992, p. 67

(i) Read the data into Minitab with the command MTB > READ 'A:\DATA07.DAT' into C1-C2, and name the columns C1 Year(X), C2 Women(Y).

(ii) Plot the data and name the type of curve. Indicate what types of transformation may be appropriate to straighten these data.

(iii) Assume that the transformation $y \rightarrow \sqrt{y}$ is appropriate. Transform the data and calculate the least squares regression line $Y = aX + b$ for the data.

4 In this exercise I would like you to examine two variables from the UNICEF data file DATA11.DAT, GNP per capita and life expectancy. GNP per capita is a measure of the wealth of a country. Life expectancy is a measure of the overall health of the people in a country and although it is not such a sensitive indicator as the perinatal or infant mortality rate it is nevertheless an important measure.

(i) Read the data into Minitab with the command MTB > READ 'A:\DATA11.DAT' into C1-C12, and name C2 GNP(X) and C11 Lifex(Y).
(ii) Plot the data and name the type of curve. Indicate the kinds of transformations that may be appropriate.
(iii) Assume that the transformation $x \rightarrow \log x$ (logarithms to the base 10) is appropriate, transform the data and calculate the least-squares regression line $Y = aX + b$. Collect the residuals from the fitted straight line and plot them against the transformed X variable.
(iv) Comment briefly on the analysis.

Answers to Exercises

1

```
MTB > READ 'A:\DATA08.DAT' into C1-C3
      5 ROWS READ

ROW    C1      C2     C3

  1   0.5    88.9    0.7
  2   1.5    78.4    6.3
  3   2.5    75.1   10.6
  4   3.5    59.9   25.6
 .  .  .
```

```
MTB > PLOT C2 against C1

        -
Und(Y) -
        -   *
        -
     80+              *
        -                       *
        -
        -
        -
     60+                                  *
        -
        -
        -
        -
     40+
        -                                                      *
        -
          --------+---------+---------+---------+---------+--Age(X)
                1.2       2.4       3.6       4.8       6.0
```

```
MTB > #This is a type 4 curve
MTB > #appropriate transformations are
MTB > #x→x², x→x³ etc. and/or
MTB > #y→y², y→y³ etc.

MTB > PLOT C3 against C1
        -
     30+
        -                                                      *
Dir(Y) -                                  *
        -
        -
     20+
        -
        -
        -
        -
     10+                          *
        -
        -              *
        -
        -
      0+    *
          --------+---------+---------+---------+---------+--Age(X)
                1.2       2.4       3.6       4.8       6.0

MTB > #This is a type 3 curve;
MTB > #appropriate transformations are
MTB > #x→x², x→x³ etc. and/or
MTB > #y→y, y→log y etc.
MTB > #For undirected anger x→x²
MTB > LET C4 = (C1)**2
MTB > #For regression line.
MTB > BRIEF=3
MTB > REGRESS Y values in C2 against 1 predictor in C4
```

```
The regression equation is
Und(Y) = 83.3 - 1.37 Age(TX)

Predictor        Coef       Stdev     t-ratio         p
Constant       83.339       3.224       25.85     0.000
Age(TX)       -1.3701      0.1867       -7.34     0.005

s = 5.413       R-sq = 94.7%     R-sq(adj) = 93.0%

Analysis of Variance

SOURCE         DF        SS         MS        F        p
Regression      1    1577.6     1577.6    53.83    0.005
Error           3      87.9       29.3
Total           4    1665.5

Obs.  Age(TX)  Und(Y)    Fit  Stdev.Fit  Residual  St.Resid
  1       0.3   88.90  83.00       3.19      5.90      1.35
  2       2.2   78.40  80.26       2.96     -1.86     -0.41
  3       6.2   75.10  74.78       2.60      0.32      0.07
  4      12.3   59.90  66.56       2.43     -6.66     -1.38
  5      36.0   36.30  34.02       5.19      2.28      1.49

MTB > #Now for the case of directed anger.
MTB > #Using x→log x.
MTB > LET C5 = LOGT(C1)
MTB > #Second regression line.
MTB > BRIEF=3
MTB > REGRESS Y values in C3 against 1 predictor
in C5

The regression equation is
Dir(Y) = 5.65 + 26.9 Age(T2X)

Predictor        Coef      Stdev     t-ratio         p
Constant        5.646      3.246        1.74     0.180
Age(T2X)       26.937      6.678        4.03     0.027

s = 5.477       R-sq = 84.4%    R-sq(adj) = 79.2%

Analysis of Variance
SOURCE         DF        SS         MS        F        p
Regression      1    488.03     488.03    16.27    0.027
Error           3     89.98      29.99
Total           4    578.01
```

```
Obs.Age(T2X)   Dir(Y)     Fit Stdev.Fit   Residual St.Resid
  1    -0.301    0.70   -2.46      4.81       3.16     1.21
  2     0.176    6.30   10.39      2.63      -4.09    -0.85
  3     0.398   10.60   16.37      2.51      -5.77    -1.18
  4     0.544   25.60   20.30      2.87       5.30     1.14
  5     0.778   28.00   26.61      3.92       1.39     0.36
```

For undirected anger the regression equation is

% undirected anger = 83.3 − 1.37(age in years)^2

For directed anger the regression equation is

% directed anger = 5.65 + 26.9(logarithm of age in years)

2

```
MTB > PLOT C2 against C1

         -
         -                                                 *
      75+
         -                                               *
Tons(Y)-
         -                                          *
         -
      50+                                      *
         -                                *
         -
         -                            *
         -
      25+                        *
         -                  *
         -             *
         -        *
         -
       ----+---------+---------+---------+---------+------Knots(X)
          10.0      12.0      14.0      16.0      18.0

MTB > #This is a type 3 curve.
MTB > #The appropriate kinds of transformation are
MTB > #x→x², x→x³ etc. and/or
MTB > #y→√y, y→log y etc.
MTB > LET C3 = SQRT(C2)
MTB > PRINT C1-C3
```

```
ROW   Knots(X)   Tons(Y)    SQRT(Y)

  1       10.0      12.2    3.49285
  2       11.0      14.9    3.86005
  3       12.0      18.1    4.25441
  4       13.0      24.2    4.91935
  5       14.0      33.8    5.81378
  6       15.0      42.5    6.51920
  7       16.0      51.7    7.19027
  8       17.0      59.9    7.73951
  9       18.0      70.1    8.37257
 10       18.5      79.7    8.92749

MTB > #Regression line.
MTB > BRIEF=2
MTB > REGRESS Y value in C3 against 1 predictor in C1

The regression equation is
SQRT(Y) = - 3.37 + 0.656 Knots(X)

Predictor      Coef        Stdev     t-ratio        p
Constant    -3.3696       0.2935      -11.48    0.000
Knots(X)    0.65596      0.01994       32.89    0.000

s = 0.1764     R-sq = 99.3%     R-sq(adj) = 99.2%

Analysis of Variance

SOURCE       DF        SS         MS          F        p
Regression    1    33.659     33.659    1081.90    0.000
Error         8     0.249      0.031
Total         9    33.908

Unusual Observations
Obs.Knots(X)   SQRT(Y)      Fit   Stdev.Fit   Residual   St.Resid
  1      10.0   3.4928   3.1899      0.1048     0.3029      2.14R

R denotes an obs. with a large st. resid.
```

The regression equation is as follows:

square root of tons of fuel oil = $-3.37 + 0.656$ (speed in knots)

3

```
MTB > READ 'A:\DATA07.DAT' into C1-C2
9 ROWS READ

ROW      C1       C2
  1    1971     94.6
  2    1976    101.9
  3    1981    128.6
  4    1983    127.4
  .   .   .
MTB > PLOT C2 against C1

     180+
        -                                                *
Women(Y)-                                            *  *  *
        -
        -                                          *
     150+                                        *
        -                                      *
        -                                    *
        -
        -                             *  *
     120+
        -
        -
        -                 *
        -        *
      90+
        -
          ----+---------+---------+---------+---------+------Year(X)
           1970.0    1975.0    1980.0    1985.0    1990.0

MTB > #This is also a type 3 curve.
MTB > #Appropriate transformations are,
MTB > #x→x², x→x³ etc. and/or
MTB > #y→√y, y→log y etc.
MTB > LET C3 = SQRT(C2)
MTB > #Regression line.
MTB > BRIEF=2
MTB > REGRESS Y value in C3 against 1 predictor in C1

The regression equation is
SQRT(Y) = - 364 + 0.189 Year(X)

Predictor       Coef      Stdev   t-ratio       p
Constant     -363.58      26.48    -13.73   0.000
Year(X)      0.18923    0.01335     14.18   0.000

s = 0.2609     R-sq = 95.3%    R-sq(adj) = 94.8%
```

```
Analysis of Variance

SOURCE            DF       SS       MS        F        p
Regression         1   13.688   13.688   201.01    0.000
Error             10    0.681    0.068
Total             11   14.369

Unusual Observations
Obs. Year(X)  SQRT(Y)     Fit Stdev.Fit  Residual  St.Resid
  1     1971   9.7263  9.3934    0.1922    0.3328    1.89 X

X denotes an obs. whose X value gives it large influence.
```

The regression equation is as follows:

square root of the no. of women in thousands
$= -364 + 0.189(\text{year})$

4

```
MTB > READ 'A:\DATA11.DAT' into C1-C12
     59 ROWS READ

ROW   C1    C2   C3   C4   C5   C6    C7   C8   C9   C10

  1    1    80    5   10   35   17   173   45   21   300
  2    1   610    6   15   34    *   173   56   29     *
  3    1   180    7   11    7   17   144    *    *   170
  4    1   120    *    *    *   34   130    *    *     *
 . . .

ROW   C11   C12

  1    48  15.7
  2    46  10.0
  3    48   8.8
  4    46  49.2
 . . .
```

```
MTB > PLOT C11 against C2
```

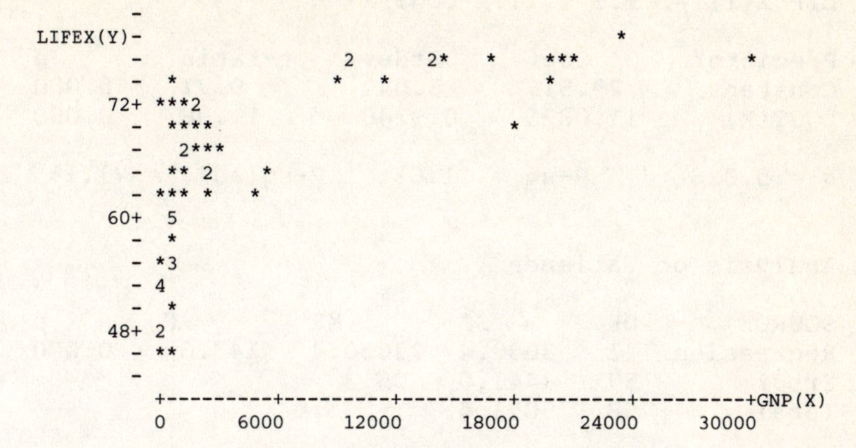

```
MTB > #This is a type 1 curve;
MTB > #Possible transformations are
MTB > #x→√x, x→log x etc. and/or
MTB > #y→y², y→y³ etc.
MTB > LET C14 = LOGT(C2)
```

```
MTB > PLOT C11 against C14
```

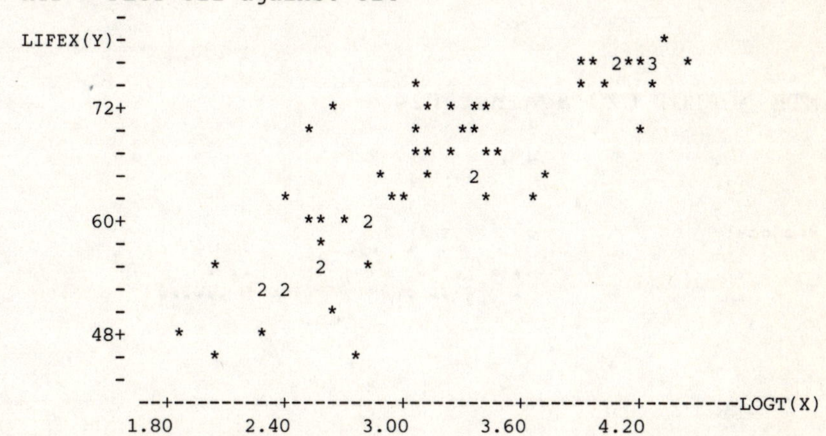

```
MTB > #Regression line.
MTB > BRIEF=2
MTB > REGRESS Y value in C11 against 1 predictor in C14;
SUBC> RESIDUALS into C20.
```

The regression equation is
LIFEX(Y) = 29.5 + 11.1 LOGT(X)

Predictor	Coef	Stdev	t-ratio	p
Constant	29.517	3.041	9.71	0.000
LOGT(X)	11.0825	0.9248	11.98	0.000

s = 5.028   R-sq = 71.6%   R-sq(adj) = 71.1%

Analysis of Variance

SOURCE	DF	SS	MS	F	p
Regression	1	3630.4	3630.4	143.60	0.000
Error	57	1441.0	25.3		
Total	58	5071.4			

Unusual Observations

Obs.	LOGT(X)	LIFEX(Y)	Fit	Stdev.Fit	Residual	St.Resid
2	2.79	46.000	60.385	0.764	-14.385	-2.89R
36	2.54	70.000	57.711	0.899	12.289	2.48R
38	2.63	71.000	58.702	0.845	12.298	2.48R
45	3.07	75.000	63.520	0.668	11.480	2.30R

R denotes an obs. with a large st. resid.

MTB > PLOT C20 against C14

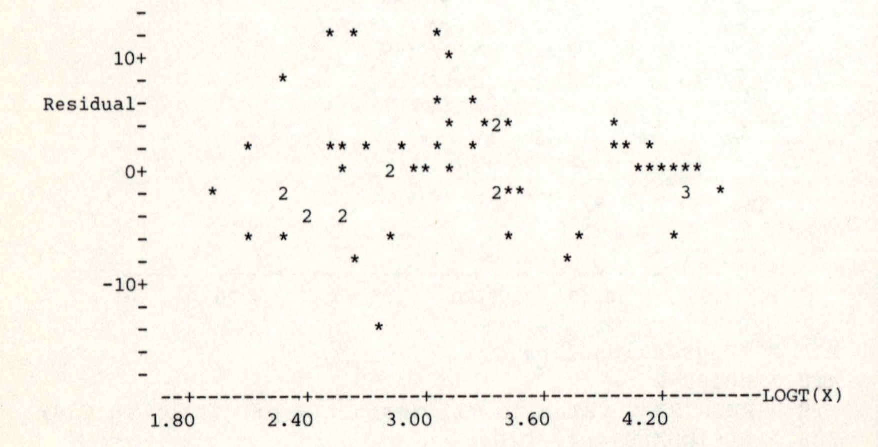

The regression equation is

life expectancy at birth in years = 29.5 + 11.1(log GNP)

(logarithms to the base 10). The R^2 of this equation is relatively high at 71% and the residuals from the fit are unpatterned and mostly small. The exceptions are Angola, China, Sri Lanka and Cuba. These require further investigation and they do serve to emphasize the general rule. Cuba and Sri Lanka (and probably China, but we have no figures) spend a relatively large amount on health, as on education.

Although life expectancy at birth is a lot less sensitive an indicator of health than infant mortality this brief analysis also reinforces our previous findings that health and wealth are very closely related in most of the countries that we have in the data set.

9 The Gaussian or Normal Distribution

In this chapter we look at the Gaussian or normal distribution. The distribution is, in reality, a family of closely related distributions which can be represented by normal curves which all have a characteristic symmetrical bell-shaped appearance.

The normal distribution has important properties relating to its shape which make it of central importance in data analysis and statistics. Close approximations to the normal curve occur not only in the natural and social world but also in probability and more importantly when we take probability samples.

As we shall see, under certain conditions repeated samples drawn from a population in a random way approach normality. Such a distribution of samples which is near normal can be modelled by the normal curve and can help us make statistical inferences about the population from the samples.

In this chapter we only look briefly at the pure or mathematical form of the distribution and then examine some close approximations to the normal. We end by using statistical inference to calculate a confidence interval for the mean of a population based on our knowledge of the mean of the sample.

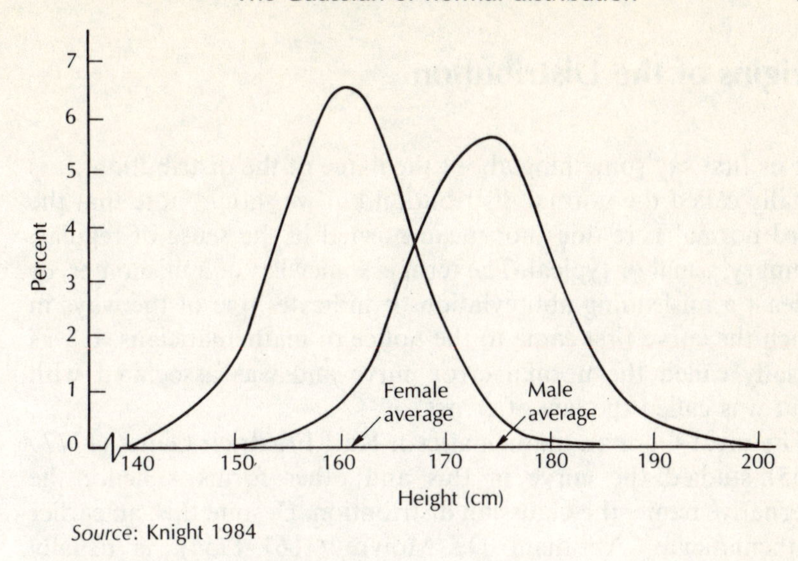

Source: Knight 1984

Figure 9.1 Two height distributions

You probably recognize the two distributions of height in figure 9.1 as normal curves. Both have an approximately normal form and represent the heights of 10,000 adults collected by Knight (1984). They have a distinct and characteristic bell shape and similar distributions are to be found in any book on statistics or data analysis and indeed, as we shall see, in many other places.

I will first say a few preliminary words about the somewhat odd name of the distribution and then look briefly at its pure or mathematical form. I will then look at three distinct ways in which approximations to the pure distribution arise, i.e.

1 spontaneously in the natural and social world;
2 as a distribution of probabilities associated with a large number of events in which each event can only happen in one of two ways, such as tossing a coin and counting the number of heads;
3 as a distribution which arises when we employ a particular type of sampling.

Origins of the Distribution

Let us first say something about the name of the distribution. It is usually called the normal distribution but we should note that the word normal here does not mean normal in the sense of regular, ordinary, usual or typical. The term is somewhat of a misnomer, or at least a misleading abbreviation. It indicates one of the ways in which the curve first came to the notice of mathematicians. It was initially called the normal error curve and was associated with what was called the law of errors.

The great German mathematician Karl Friedrich Gauss (1777–1855) studied the curve in this and other forms – hence the alternative name, the Gaussian distribution. Despite this, an earlier mathematician, Abraham De Moivre (1667–1754), is usually credited with the discovery or at least the first published formulation of the distribution.

De Moivre was a French Protestant refugee from religious persecution who lived most of his adult life in England and was a confidant and friend of Isaac Newton. He discovered the curve when dealing with what are known as binomial expansions such as $Y = (a + x)^n$. However, De Moivre, Gauss and others also looked at errors and the form or distribution of these errors when they arise from repeated measurements with scientific instruments. With the growth of science in the eighteenth and nineteenth centuries and the development of accurate measuring instruments such as telescopes, sextants, callipers, gauges, weighing balances, clocks and so on, it was found that despite the increasing sophistication of the instruments and the use of finely graduated vernier and micrometer-screw gauges, repeated measurements produced different observed readings.

If repeated measurements were taken of a constant or 'stationary' phenomenon, say the horizontal angle between two mountain peaks from a given point, or any simultaneous measurements were made of other phenomena such as the altitude or position of a star at a particular moment in time, then these sets of observations would be clustered around the true value in a characteristic way. Some would over-read the true value while others would under-read it. The set of readings would form what came to be called a normal error curve. That is, the set of measurements was distributed about the true value in a constant and characteristic way. The more observations that are taken in such circumstances the closer will the resulting data come to the symmetrical mathematical form we now call the normal or Gaussian distribution.

The Probability Density Function

The formula, i.e. the probability density function (PDF), for the normal distribution may be given in the following form:

$$Y = \text{a function of } X = \frac{1}{\sigma\sqrt{2\pi}} \exp\left[-0.5\left(\frac{X-\mu}{\sigma}\right)^2\right]$$

X is the variable, or more properly the continuous random variable, which can take on any value between $\pm\infty$. The normal distribution Y is a function of this variable X as indicated by the expression on the right of the formula. However, do not be put off by the apparent complexity of this formula as we shall not, in this book, work directly from the formula but use instead the various tabulated values of the normal distribution to be found in Neave and other statistical tables.

It is worth noting, however, that some expressions in the above formula are familiar to us from our earlier work on standardization: exp stands for the Euler or exponential number. It is a constant and has the value 2.7182...; π is of course another constant and has the value 3.1416.... The Euler number, which is named after the Swiss mathematician Leonhard Euler (pronounced Oiler), is raised to the power $-0.5[(X - \mu)/\sigma]^2$.

X is of course the variable that makes up the distribution while μ and σ are respectively the mean and standard deviation of the distribution. You will recognize the expression $(X - \mu)/\sigma$ as the standardization formula that we used in an earlier chapter.

We should note that this is not the formula for one curve but is a generic formula for a family of curves or distributions of the normal form. That is, if we take values of X between $+\infty$ and $-\infty$ we can generate a very large, indeed infinite, number of normal curves by using different values of the mean μ and the standard deviation σ. However, each pair of values (μ and σ) that we choose can generate one and only one normal curve. Once we have specified these two parameters we have specified the curve completely. Or, to put it another way, if we have an empirical distribution of data, say heights, which looks approximately normal in shape, we can generate a number of unique normal curves using various paired values of the mean μ and standard deviation σ in the formula and see whether any of them give a good model or fit.

The Standard Normal Distribution

There is one particular pair of values of μ and σ which give rise to a special type of normal distribution, the standard normal distribution. In chapter 5 we saw that we could standardize or transform data using the general formula

$$\text{standardized score} = \frac{\text{value of the variable} - \text{level}}{\text{spread}}$$

For a standardized score using the mean and standard deviation as measures of level and spread respectively we get

$$\text{standardized score} = \frac{X - \mu}{\sigma}$$

As we saw previously, data which is standardized or transformed in this fashion will give rise to another more manageable distribution which has a mean μ of zero and a standard deviation σ of unity. So given $Z = (X - \mu)/\sigma$ we have a special case within the family of normal curves where the variable is normally distributed with a mean of zero and a standard deviation of unity. The new curve is of course still normal in form but the new variable is Z rather than the original X. As $\mu = 0$ and $\sigma = 1$ the original formula is now reduced to the following function of Z:

$$Y = \text{function of } Z = \frac{1}{\sqrt{2\pi}} \exp(-0.5\, Z^2)$$

The variable Z is called the standard normal variable and its distribution is called the standard normal distribution. It has a mean of zero and a standard deviation of unity and various values associated with this curve are to be found tabulated in statistical tables. The ordinate (height of the curve) and area under the curve (but above the Z axis) are given for various values of Z on pages 371–2 in appendix 2.

It is worth noting here that we have standardized the generic formula for a normal curve and we can of course standardize any one of the myriad normal distributions in the family in the same way to end up with a standard normal curve. Standardization is a very useful transformation which allows us to use standard normal tables for all normal curves and their approximations.

We now review some of the characteristics of the normal distribution.

As we saw earlier, all normal curves are uniquely determined by their mean and standard deviation. That is, for any given pair of values of the mean and standard deviation we can use the formula to draw one, and only one, normal curve. Each of these curves will be different but will be stamped by the common family characteristics. Each member of the normal family of curves will be unimodal and symmetrically bell shaped about the mean. The mean, mode and median are coincident.

As we know, the mean determines the position or level on the horizontal variable axis and the standard deviation governs the spread or dispersion about the mean. The curve may be 'fat' or 'thin' depending on whether the standard deviation about the mean is large or small. The tails of the curve come gently down to the variable axis and almost touch it at 3 standard deviations either side of the mean. From here the tails of the distribution extend, in theory, in each direction to infinity, and the curve comes closer and closer to the variable axis as we move away from the mean. This is sometimes described as the curve approaching a tangent with the variable axis at both plus and minus infinity. The curve is asymptotic to the horizontal axis. However, for our purposes, and indeed for most practical purposes, almost all the area under the curve and above the variable axis is contained within a relatively short distance from the mean.

We can illustrate some of these general features by looking at a sketch of the standard normal distribution (figure 9.2). As we noted above, standardization has reduced the mean of this curve to

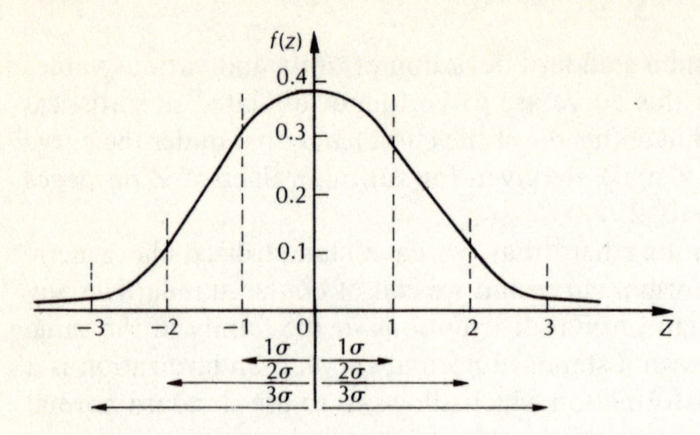

Figure 9.2 The standard normal curve: N ($\mu = 0, \sigma = 1$)

zero and the standard deviation to unity:

$$Y = \text{function of } Z = \frac{1}{\sqrt{2\pi}}\exp(-0.5Z^2)$$

It can be proved, although we will not do so here, that the area under the curve and above the variable axis for this form of the distribution is equal to one square unit. As we shall see later, this area and proportions of it form a measure of various probabilities. Of this unit area about

68% is contained between $+1\sigma$ and -1σ and the horizontal Z axis,
95% is contained between $+2\sigma$ and -2σ and the horizontal Z axis,
99.7% is contained between $+3\sigma$ and -3σ and the horizontal Z axis.

Although we have noted that the tails of the curve extend to infinity, for most practical purposes we can consider 99.7% of the distribution to be equal to unity. It is useful to remember the areas corresponding to the above six Z values; however, the areas (probabilities) corresponding to any given Z value can be found in statistical tables such as Neave. We look at how to use these tables in the next chapter.

Let us now turn from the conceptual world of mathematics and our description of the precise form of the normal distribution to some of the many empirical approximations to be found in the 'real' or everyday world.

Approximations to the Normal Distribution

The distribution of adult human height and weight

As we noted at the beginning of this chapter, a large number of adult heights collected at random, when plotted, will give a very good approximation to a normal curve.

Adolphe Quetelet (1796–1874), a Belgian mathematician, astronomer, sociologist, early statistician and general polymath, was one of the first people to note and describe the way that a large number of heights, weights and similar measurements are distributed in approximately normal form. For instance, in one study he used the measured chest girth of 5738 Scottish soldiers which varied between 33 and 46 inches. The mean was found to be about 40 inches and he found that the empirical distribution of the measurements could be closely modelled by a normal curve. He found that anthropometric and other naturally occurring data could be studied using the normal curve as a model and he was even convinced that some social events, such as crime and conviction rates, showed a distinct pattern which could be modelled by the normal and other mathematical curves.

The data that Ian Knight (1984) used to draw the two curves shown at the beginning of this chapter were the measured heights of just less than 10,000 adults in Great Britain. The levels and spread were as follows:

Men's data

n	4702
Mean	173.9 cm (5″ 8.5″)
Median	173.8 cm
Standard deviation	9.6 cm

Women's data

n	5,158
Mean	160.9 cm (5″ 3.5″)
Median	160.8 cm
Standard deviation	7.9 cm

Both sets of data are closely normal in form, but of course they are still approximations to the mathematically defined curves which are continuous rather than made up of discrete elements such as we have here. Both of these near-normal data distributions, as we shall see, can be closely modelled by normal curves defined by their means and standard deviations. The mean and standard deviation are referred to as the defining parameters of the curves.

It is worth noting that for both male and female distributions the two measures of average, the mean and the median, are very close together, being in each case only 0.1 cm apart. This is further confirmation of our view that the curves are near normal in form. A normal curve is of course symmetrical about the mean so that the three measures of level or average, the mean, mode and median, coincide.

Let us now turn to a second way in which we can get distributions which are near normal in shape.

Binomial probabilities

The kind of probabilities that we are talking about here are those which arise from activities such as coin tossing. We are dealing with finite or discrete happenings and we can approach the probabilities involved by simply counting the events. In the case of repeated coin tossing we are dealing with a collection of events where each separate event can happen in one of two ways. These are sometimes called Bernoulli events. We might be interested in the probability of getting, say, 3 heads in the tossing of 5 coins, where in each toss we can only get either a head or a tail. The individual head or tail outcome is referred to as a two-state event, i.e. either one event or the other always happens. A series or collection of such two-state Bernoulli events will give us a Binomial distribution. It is important to note here that each of these two-state events must be independent, i.e. the outcome of one event must not affect the outcome of the next or subsequent events. In other words the probability, in our case one-half, must remain the same between events.

When we consider a large number of such events we find that the resulting binomial distribution can be closely modelled by a normal curve. Before looking at some of these binomial distributions let us say a few words about probability.

Probability or chance is about the uncertain outcome of events. It is often associated in our minds not only with coin tossing but also with dice, roulette wheels, and games of chance and gambling in general. In fact much probability theory in mathematics did have its origins in gambling. Only relatively recently, say in the last 200 or so years, has probability been associated with more respectable pursuits and disciplines such as demography, social sciences and statistics.

Gamblers wish to win, and gamblers who were also mathematicians were no different in this respect. Indeed Fibonacci of Pisa (1170–1250) only seemed to be interested in the mathematics if it could produce an increase in winnings. However, despite the materialistic thrust of their enquiry he and others noticed that over many chance events a pattern or distribution of success and failure seemed to emerge with a high degree of consistency. Few bookmakers are mathematicians and yet their appreciation of these selfsame patterns that arise from random or chance events enables them to make precise and profitable calculations of the odds over a large number of events.

Our interest here is in binomial distributions arising from coin tossing but mathematicians also noted that a great number of natural phenomena in the world as we know it also appear to be distributed by chance. When we are dealing with collections of two-state events there are two points that we need to note: (a) individual occurrences cannot be predicted, but (b) in the long term or when data are aggregated, predictable and stable patterns emerge which can be measured with a high degree of accuracy. These patterns are called probability distributions. Both the normal and the binomial distributions are examples of such probability distributions. What is unpredictable and uncertain at an individual level is both patterned and predictable at a higher level. You may be familiar with this type of distinction between individual and aggregated qualities from other areas of social science. For instance, we know that in women giving birth in Britain factors such as low social class (Registrar General Classes IV and V), heavy smoking, short stature, living in the north of Britain and extreme age (either very young or very old in childbearing terms) are consistently associated with patterns of low birth-weight and high infant mortality rates. However, these known patterns or distributions cannot tell us anything precisely about the outcome of any *one woman's* confinement. They can of course give us an idea of the probability of risk.

We should also note that in dealing with births we have not only the distinction between individual and aggregated events but the individual events may be of the two-state type mentioned earlier. In distributions of birth and mortality, the baby can be, for instance, either male or female, and either dead or alive. Similarly a person may or may not have a disease or health condition. The outcome of testing a new drug may be measured starkly in terms of success or failure. Sets or collections of two-state events such as these where something does or does not happen, exists or does not exist, can all give rise to probability distributions. Such probability distributions are relatively common in the natural and social world.

Let us now return to our coin tossing which can be used to model such events and distributions. We will look at some elementary ideas of probability in relation to the binomial distribution.

There are various different ways of looking at probability, but we will start with a definition of theoretical or *a priori* probability. This definition of probability is readily grasped but it is by no means free of ambiguity (the phrase 'equally likely' embedded in it is suspect) but it will suffice for our purposes here.

Suppose an event E can happen in f ways out of a total of n theoretically possible equally likely ways. Then the probability of the event occurring is given by

$$\text{probability of event} = P(\text{E}) = \frac{f}{n}$$

Consider for example the probability of getting a 4 in a single toss of a fair die. In this case we get

$$\text{probability of event} = P\,(\text{getting a 4}) = \frac{f}{n} = \frac{1}{6}$$

Also it can be seen that the probability of not getting a 4 in a single toss is

$$\text{probability of event} = P\,(\textit{not}\text{ getting a 4}) = \frac{5}{6}$$

i.e. $1-1/6=5/6$. From the way it is defined we can see that probabilities always lie on a scale between 0 and 1 inclusive. The probability, P, of an event always lies between 0 and 1, i.e.

$$\text{probability of event} = P(\text{E}) \qquad 0 \leqslant P(\text{E}) \leqslant 1$$

In the above inequality 0 indicates the certainty that the event will not happen and at the other end of the scale 1 indicates the certainty that the event will happen. Between these two extremes we have a range of fractional probabilities of the event happening or not happening. ($\leqslant$ means less than or equal to). The nearer the probability is to 1 then the more likely it is that the event happens. For instance, an event with a probability of 0.95 could be expected to happen in the long run in 95% of cases.

The sum of the probabilities of an event happening and not happening must be equal to 1 in the two-state systems that we are considering. For example, as above, the probability of getting a 4 and not getting a 4 with a single throw of a fair die must be equal to 1.

$$P(\text{getting a 4}) + P(\text{not getting a 4}) = \frac{1}{6} + \frac{5}{6} = \frac{6}{6} = 1$$

Let us now consider the pattern of outcomes of tossing two, three, four and more fair coins. A fair coin is one which is likely, in the long run, to give an equal number of heads and tails. Also let us imagine that we are interested in the number of heads that result from the various tossings. That is, the number of heads is the variable (the thing that varies) that we are interested in. We will call this variable X. From our original definition, the proportion of times each value of the variable occurs is called the relative frequency or probability of the event.

Tossing two coins (or tossing one coin twice) Suppose a fair coin is tossed twice. There are four equally likely outcomes, i.e. HH, HT, TH, and TT. The variable X, is the number of heads and we can see that only certain outcomes are possible. It is only possible to get zero heads, one head or two heads. That is, the variable X can only take on one of the three values, 0, 1 or 2. The probabilities associated with these values of the variable add up to 1. That is,

$$P\,(\text{no heads}) + P\,(\text{one head}) + P\,(\text{two heads}) = 1$$
$$P\,(X = 0) + P\,(X = 1) + P\,(X = 2) = 1$$

As there are four possible outcomes in total the value of each of the probabilities is

$$P\,(\text{no heads}) = P\,(X = 0) = 1/4 = 0.25$$
$$P\,(\text{one head}) = P\,(X = 1) = 2/4 = 0.5$$
$$P\,(\text{two heads}) = P\,(X = 2) = 1/4 = 0.25$$

The above probabilities can be seen to form a probability distribution for the random or chance variable X (getting a head) and as such can be illustrated by a simple histogram (figure 9.3).

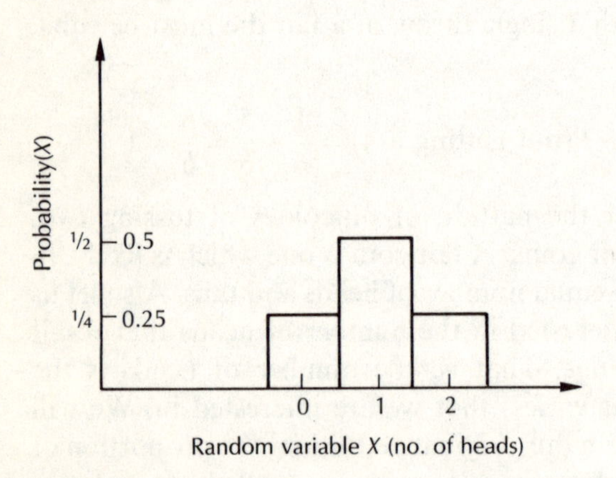

Figure 9.3 Tossing two coins

Tossing three coins (or tossing one coin three times) If a similar tossing of three fair coins is considered the following probability distribution occurs. This time there are eight possible outcomes, i.e. HHH, HHT, HTH, THH, TTH, THT, HTT and TTT. We will again take the random variable X to be the number of heads. This variable X can only take on the values 0, 1, 2 or 3 and the probabilities associated with these values of the variable can be found by counting to be as follows:

P (no heads) $= P(X = 0) = 1/8 = 0.125$
P (one head) $= P(X = 1) = 3/8 = 0.375$
P (two heads) $= P(X = 2) = 3/8 = 0.375$
P (three heads) $= P(X = 3) = 1/8 = 0.125$

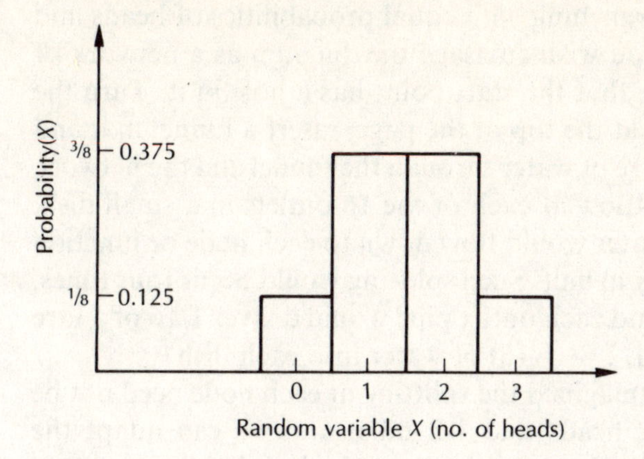

Figure 9.4 Tossing three coins

As before this distribution of probabilities can be illustrated by a histogram (figure 9.4).

Tossing four or more coins The pattern that emerges when we consider the tosses of four or more fair coins can be found in a similar way but we probably need a more orderly way of laying down the total possibilities. This can be done with a simple

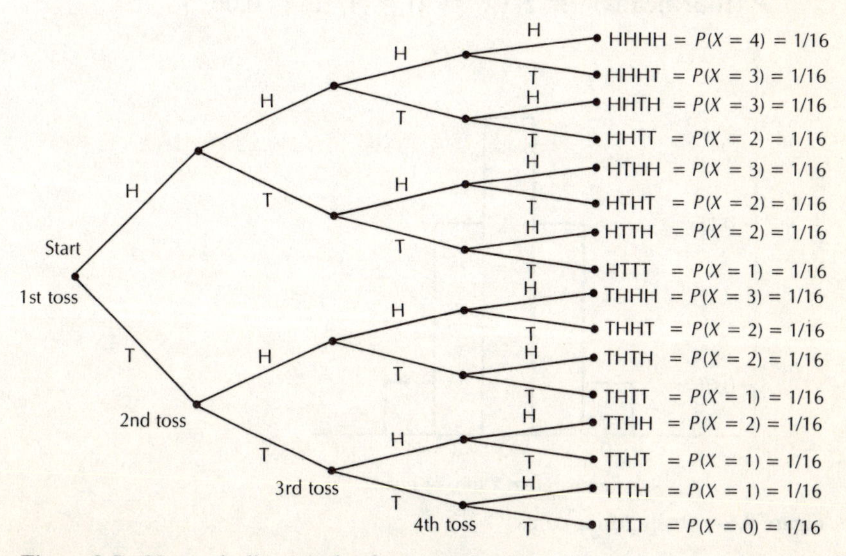

Figure 9.5 Network diagram for four tosses

network or probability tree. Let us consider the possible outcome of four tosses (figure 9.5).

The sequences of heads and tails can be found by following the diagram from the start through the four tosses. At each node or junction there is a branching with equal probabilities of heads and tails. You could if you wish envisage the diagram as a network of small pipes. Imagine that the start point has a hole in it. Turn the start point until it is at the top of the page, insert a funnel in it and imagine pouring a litre of water through the funnel and the network and catching the outflow at each of the 16 outlets in a small dish. Under gravity the water would flow down to each node or junction and then split exactly in half. Such splitting would occur four times, once for each toss, and each outlet pipe would deliver 1/16 of a litre ($1/2 \times 1/2 \times 1/2 \times 1/2 = 1/16$) of water into each dish.

As can readily be imagined the splitting at each node need not be equal as above (1/2 heads and 1/2 tails) and we can adapt the system to splitting the flow, i.e. the individual probabilities, to say 1/4 and 3/4, or 1/3 and 2/3. Such unequal probabilities will still be binomial in form. However, to come back to our equal probabilities, i.e. the outcomes of a fair coin, by counting as before we can get the following probability distribution.

$$
\begin{aligned}
P\,(\text{no heads}) &= P\,(X = 0) = 1/16 = 0.0625\\
P\,(\text{one head}) &= P\,(X = 1) = 4/16 = 0.25\\
P\,(\text{two heads}) &= P\,(X = 2) = 6/16 = 0.375\\
P\,(\text{three heads}) &= P\,(X = 3) = 4/16 = 0.25\\
P\,(\text{four heads}) &= P\,(X = 4) = 1/16 = 0.0625
\end{aligned}
$$

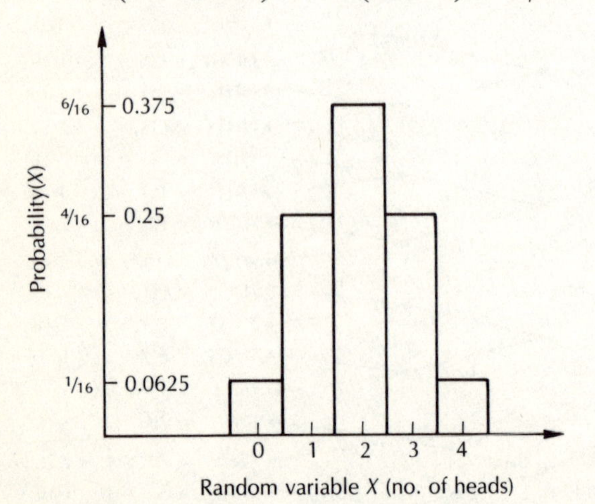

Figure 9.6 Tossing four coins

As before this distribution of probabilities can be illustrated by a histogram (figure 9.6).

We could continue tossing five, six and more coins in a similar fashion and counting the number of heads that result but it would be somewhat tedious. Instead let me illustrate the probability distribution associated with tossing six coins and then say something about the general form the distribution takes as the number of tosses gets large.

The probability distribution that arises from six tosses is as follows (figure 9.7):

P (no heads) $= P(X = 0) = 1/64 = 0.0156$
P (one head) $= P(X = 1) = 6/64 = 0.0940$
P (two heads) $= P(X = 2) = 15/64 = 0.2344$
P (three heads) $= P(X = 3) = 20/64 = 0.3125$
P (four heads) $= P(X = 4) = 15/64 = 0.2344$
P (five heads) $= P(X = 5) = 6/64 = 0.0940$
P (six heads) $= P(X = 6) = 1/64 = 0.0156$

The asterisks mark the middle of the top of the histogram blocks.

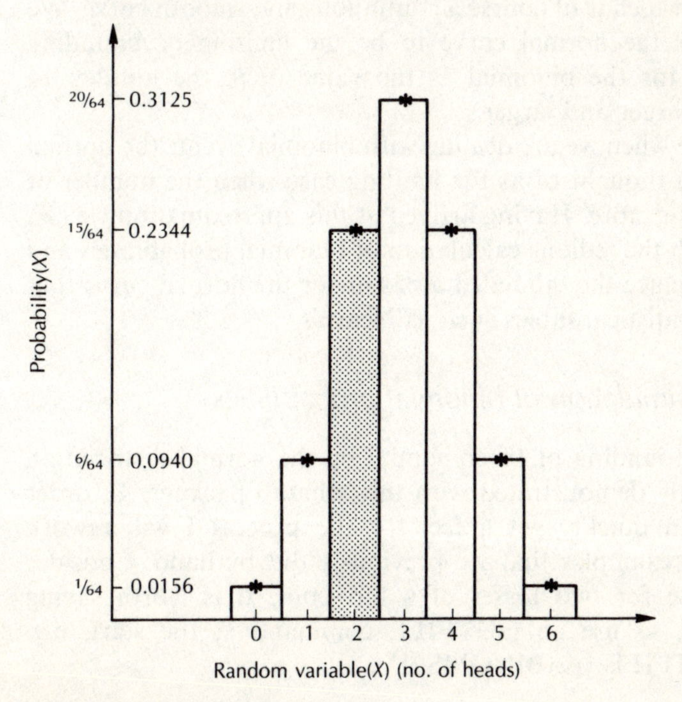

Figure 9.7 Tossing six coins

The probabilities associated with each outcome, no heads, one head, two heads and so on, are given by the height of the respective columns of the histogram in figure 9.7 as each column is of equal width. If we take each column to be 1 unit wide we can see that the area of each column represents the associated probability. For instance, the column representing the probability of two heads is 15/64 units high and 1 unit wide and so has an area of $15/64 \times 1 = 15/64$ square units, which I have shaded. If we consider all the probabilities in this fashion we can see that the total area enclosed by all the histogram columns is $64/64 = 1$ square unit.

It is not difficult to imagine that if we were dealing with a large number of tosses, say over 30, then, as the area representing the sum of the probabilities would still be equal to one square unit, the individual histogram columns would get thinner and thinner. We can also imagine that in this case the upper 'stepped' boundary of the histogram would approach a type of curve, albeit a rather jagged curve. Despite the fact that the binomial distribution is discrete and its probabilities vary in discrete pieces or 'jumps', we can say that as the number of tosses n gets larger and larger the curve can be more and more closely approximated by a normal distribution which is of course a continuous and smooth curve. We can consider the normal curve to be the limiting or bounding distribution for the binomial as the value of n, the number of tosses, gets larger and larger.

In practice when we are dealing with binomial events the normal curve can be thought of as the limiting case when the number of events is 30 or more. Having arrived at this approximation we can dispense with the tedious calculation of binomial probabilities and use in their place the tabulated areas under the normal curve that we find in statistical tables such as Neave's.

Computer simulation of binomial probabilities

The above bounding of the binomial by the normal distribution can be readily demonstrated with the Minitab package. In order that you can quickly get a feel for the process I will rework some of the examples that we previously did by hand. Consider first the case for two tosses of a fair coin. It is worth saving your output so use an OUTFILE command at the start, e.g. MTB > OUTFILE 'A:\BINOMS01'.

The appropriate Minitab commands are

```
MTB > PDF;
SUBC> BINOMIAL N=2 P=.5.
```

You should follow these examples through on the computer. Please note the punctuation in the above command. The command PDF stands for probability density function and in this case the PDF that we require is the probability density function of the binomial where n, the number of events, is 2, and the probability associated with each individual event is 1/2. With these commands you should get the following print-out.

```
BINOMIAL WITH N = 2 P = 0.500000
       K          P( X = K)
       0            0.2500
       1            0.5000
       2            0.2500
```

This corresponds closely to what we obtained previously by hand for the tossing of two coins. Comparing this with what we had before (p262) you will see that K indicates the values that the variable X can take, and the P column gives the probability of obtaining those values. For example the probability of getting no heads is

$$P\,(\text{no heads}) = P\,(X = 0) = 1/4 = 0.2500$$

Similarly commands such as

```
MTB > PDF;
SUBC>BINOMIAL N=3 P=.5.
```

will produce the binomial distribution for tossing three coins in very much the form that we had previously (p263). However, it is more useful to organize the Minitab output rather differently.

Consider six tosses of a fair coin, i.e. $N = 6$ and $P = 1/2 = 0.5$. We know that we can only get the following outcomes or number of heads, 0, 1, 2, 3, 4, 5 and 6. No other values of the variable X are possible. If we put these values that the variable X can take into a column we can then match the individually calculated probabilities to them, put them in another column and then plot the PDF (the distribution) as follows:

```
MTB > SET C1
DATA> 0 1 2 3 4 5 6
DATA> END
MTB > PDF C1 and put into C2;
SUBC> BINOMIAL N=6 P=.5.
MTB > #Go into the Editor and name C1 No.Heads
MTB > #and C2 Rel.Freq.
MTB > PRINT C1 C2

ROW  No.Heads   Rel.Freq

  1         0   0.015625
  2         1   0.093750
  3         2   0.234375
  4         3   0.312500
  5         4   0.234375
  6         5   0.093750
  7         6   0.015625


MTB > PLOT C2 against C1
MTB > #You should note the order here, we wish C2
MTB > #to be the horizontal axis.
```

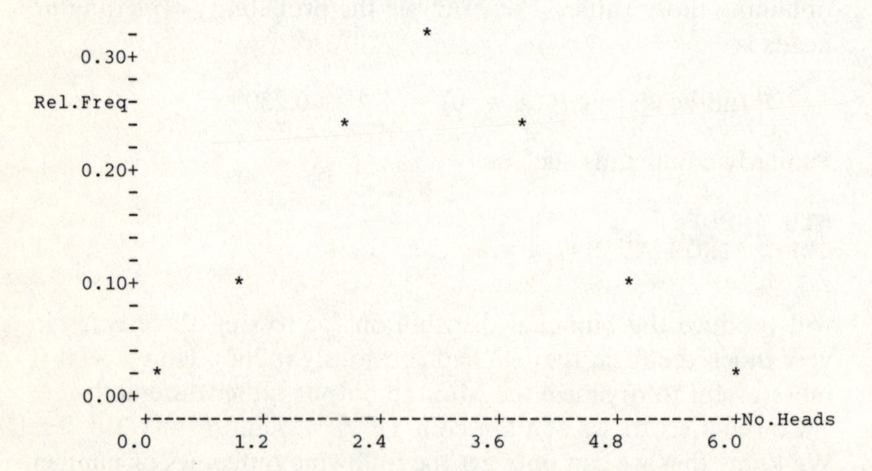

You should compare this Minitab diagram with the previous histogram for six tosses (figure 9.7, p265). What we have here, if we imagine the points to be joined up, is a frequency polygon. The asterisks represent the points at the middle of the top of the histogram blocks, but Minitab does not show the blocks themselves.

Using the same general format you can convince yourself that, as the number of tosses N becomes large, the binomial distribution of probabilities does indeed approach a normal shape. Try and illustrate 10 tosses with the following set of commands,

```
MTB > ERASE C1 C2 #To start with a clear data screen.
MTB > SET C1
DATA> 0:10 #Gives required sequence 0 1 2 .... 10.
DATA> END
MTB > PDF C1 and put into C2;
SUBC> BINOMIAL N=10 P=.5.
MTB > #Name the columns as before in the Editor.
MTB > PRINT C1 C2

 ROW   No.Heads   Rel.Freq

   1          0   0.000977
   2          1   0.009766
   3          2   0.043945
   4          3   0.117188
   5          4   0.205078
   6          5   0.246094
   7          6   0.205078
   8          7   0.117188
   9          8   0.043945
  10          9   0.009766
  11         10   0.000977

MTB > PLOT C2 against C1
```

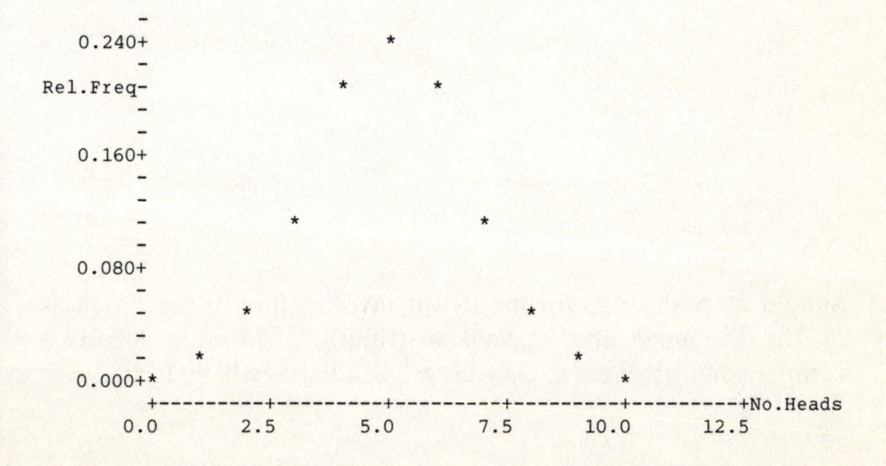

As you can see, even with $N = 10$ we are starting to get the distinctive bell shape of the normal distribution. You should now try giving the commands that are appropriate for 30 and then say 100 tosses. (Set one column with 0:30 or 0:100 respectively.)

When plotting such distributions we should remember that as N gets larger the binomial distribution approaches the normal curve more and more closely. As we know, the normal curve has tails which extend in theory to plus and minus infinity, and for large N the binomial distribution has similar tails. When plotting Binomial distributions for a large number of tosses N, we get a clearer picture of the plot by cutting off part of these tails. To do this we restrict the X axis output. Consider the following for a binomial distribution arising from 100 tosses.

```
MTB > ERASE C1 C2 #To clear C1 C2
MTB > SET C1
DATA> 0:100
DATA> END
MTB > PDF C1 and put into C2;
SUBC> BINOMIAL N=100 P=.5.
MTB > #Name the columns as before.
MTB > PLOT C2 against C1;
SUBC> XSTART at 35 and end at 65.
```

```
         -
  0.090+
         -
Rel.Freq-                  ** *
         -                *    *
         -              *        *
  0.060+              *           *
         -
         -           *              *
         -
         -           *               *
  0.030+           *                  *
         -          *                   *
         -         *                     *
         -       *                        *
         -    * *                           * *
  0.000+ ** *                                  ** **
         +---------+---------+---------+---------+----No.Heads
      35.0      42.0      49.0      56.0      63.0
      69 Points out of bounds
```

Should we wish to go further in our investigation of the closeness of the binomial and normal distributions Minitab has two comparisons or checks. One uses NSCORES which I shall not and the other utilizes a double plotting technique MPLOT which we will look at briefly before going on to the third way in which we can get an approximation to the normal distribution.

MPLOTS to Check Normality

We have seen how Minitab can produce probability distributions (probability density functions) for a binomial series of two-state events once we specify the parameters N (the number of tosses) and P (the probability). It can also produce probability distributions for other statistical functions such as uniform, Poisson, χ^2 etc. and of course the normal.

Here we will produce both a binomial and a matching or fitted normal distribution (PDFs) and by plotting them on the same axis get a visual check of the closeness of fit between the binomial and the normal curves.

The parameters of the binomial distribution are, as we have seen, the number of events N and the probability P for each individual outcome. For the Normal distribution the parameters are the mean and the standard deviation. In Minitab these are referred to as MU and SIGMA respectively.

When the number of events N becomes relatively large we can calculate the parameters of a normal curve that will give a reasonable fit to a binomial distribution as follows:

Mean of the normal distribution = MU = NP
Standard deviation of the normal distribution =
SIGMA = $\sqrt{[NP(1-P)]}$

Let us now illustrate this fitting process by looking at the binomial distribution for 10 tosses, i.e. $N=10$ and $P=0.5$, and then constructing the fitted normal curve with mean and standard deviation as follows:

MU = NP = $10 \times 0.5 = 5$
SIGMA = $\sqrt{[NP\,(1-P)]} = \sqrt{[10 \times 0.5 \times 0.5]} = \sqrt{2.5} =$
1.58

Our small program of commands to produce the plots is as follows. (As with the other exercises and examples you should try this out for yourself rather than just read it through.)

```
MTB > SET C1
DATA> 0:10
DATA> END
MTB > PDF C1 and put into C2;
SUBC> BINOMIAL N=10 P=.5.
MTB > #This puts the Binomial distribution of
MTB > #probabilities (PDF) into C2.
MTB > PDF C1 and put into C3;
SUBC> NORMAL MU=5 SIGMA=1.58.
MTB > #This puts the Normal distribution of
MTB > #probabilities(Mean 5,Stan.Devn 1.58)
MTB > #into C3.
MTB > PRINT C1-C3

ROW  No.Evnts    Bin.Prob    Nor.Prob

  1         0    0.000977    0.001689
  2         1    0.009766    0.010245
  3         2    0.043945    0.041629
  4         3    0.117188    0.113323
  5         4    0.205078    0.206666
  6         5    0.246094    0.252495
  7         6    0.205078    0.206666
  8         7    0.117188    0.113323
  9         8    0.043945    0.041629
 10         9    0.009766    0.010245
 11        10    0.000977    0.001689

MTB > MPLOT C2 against C1 and C3 against C1

       -                         B
0.240+                           A
       -
       -                    2         2
       -
       -
0.160+
       -
       -
       -              2                   2
       -
0.080+
       -
       -         2                            2
       -
       -     2                                    2
0.000+ 2                                              2
       +---------+---------+---------+---------+---------+
      0.0       2.5       5.0       7.5      10.0      12.5
   A = C2 vs. C1                  B = C3 vs. C1
```

In the above graph the binomial distribution in C2 has been plotted with As and the Normal distribution in C3 with Bs, and as you will remember where the two values coincide a 2 is printed by Minitab. As you can see, even when the number of binomial events *N*, is as low as 10 we have been able to fit a normal curve very closely to the binomial.

What we have done in this section of the chapter is to demonstrate that the binomial distribution approaches the normal distribution as the number of events *N*, increases.

Before leaving this topic of curve fitting and simulation I would like you to try to fit two normal curves to Ian Knight's data on adult heights with which we started the chapter. You will remember that we claimed in the early part of the chapter that such sets of heights approximated closely to normal distributions. If this is indeed the case, given the basic parameters of the two distributions, we should be able to simulate these empirical distributions with normal ones generated in Minitab.

We use a base of 140 to 200 cm which should enable us to simulate most of the two distributions. If you look back to page 257 you will see that the parameters for the two empirical distributions are as follows:

Men	mean = 173.9 cm	standard deviation = 9.6 cm
Women	mean = 160.9 cm	standard deviation = 7.9 cm

The following commands should give you the graph of the two distributions on the same axis:

```
MTB > SET C1
DATA> 140:200
DATA> END
MTB > PDF C1 and put into C2;
SUBC> NORMAL MU=173.9 SIGMA=9.6.
MTB > PDF C1 and put into C3;
SUBC> NORMAL MU=160.9 SIGMA=7.9.
MTB > MPLOT C2 against C1 and C3 against C1
```

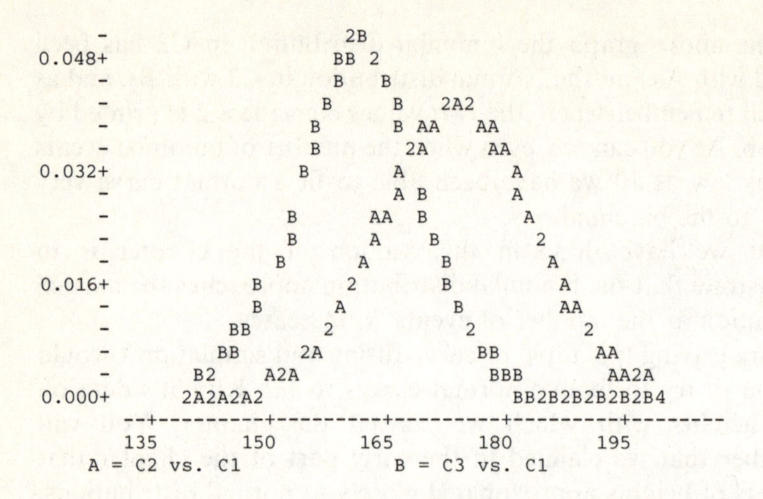

As can be seen, the normal simulations using the above parameters give us distributions which seem, despite Minitab's restricted graphics capability, to follow closely the two curves drawn from Knight's empirical data with which we started this chapter.

Let us now turn to the third and last way we shall consider for getting a normal distribution. This method also considers the normal curve as the limiting form of a probability distribution. The probability distribution is what is called the sampling distribution of means and it arises not from coin tossing and the like but from a particular type of sampling technique, i.e. random sampling.

The sampling distribution of means

This approximation to the normal curve also arises from probability distributions which are in many ways similar to those that we have just looked at that are associated with coin tossing. However, here the element of probability or uncertainty is ensured by the type of sampling employed, random sampling.

Contrary to the way we tend to use these two words random and sample in everyday life, in mathematics they have a defined and precise meaning. In the vernacular a random choice, say of beers in a pub, has a loose meaning of choosing a thing in a haphazard, unsystematic or purposeless way. This is very far from the mathematical or statistical use of the term.

Sampling is a way of choosing a set or group of elements from the total set or population of such elements under consideration. The population may consist of, say, heights, weights or IQ scores, and the sample will form a subset of such elements. Random sampling is a process in which each unit or element in the set or population has the same chance or probability of being chosen for the sample. Such a process ensures that we end up with a probability distribution. When we have such a random sample a theorem called the central limit theorem enables us to apply the laws of probability to bridge the gap between the sample and the population. That is we can use what we know about the sample, say our knowledge of its mean or median, to infer things about the population mean or median. This is called statistical inference. At the heart of such a process is the fundamental central limit theorem. A study of this theorem lies outside the scope of this introduction to data analysis but we shall see that its use leads us back to the normal distributions which have been the focus of our attention in the present chapter.

For an effective inference we want a sample which provides us with a representation of the population. That is, we need in some way to match or relate the sample characteristics systematically to the characteristics of the parent population from which it was drawn.

Let us imagine that we have a large population, say a population of heights X in centimetres, and that the mean height of this population is μ and its standard deviation is σ. If we now take a random sample of say 30 heights (N = 30) from the population we can calculate the mean of this sample by, for instance, adding up the 30 heights and then dividing the sum by 30. Let us call the mean of this first sample $\bar{X}_1$. If we now repeat the process and take many such random samples of size 30 and find their means we can label them $\bar{X}_2$, $\bar{X}_3$, $\bar{X}_4$, ..., $\bar{X}_r$. These r samples now form a probability distribution that is usually called the sampling distribution of means.

If we take many such samples the central limit theorem tells us that the shape of the sampling distribution of means approaches the normal distribution as r, the number of samples, increases. We might intuitively expect this to be the case if the parent population being sampled is itself a normal distribution; however, surprisingly this result holds good for many distributions which are far from normal in shape as we shall see later. We shall take a uniform distribution which has a very uncompromising block shape and from it produce by the above process of random sampling a sampling distribution of means which is very close to normal in shape.

Apart from randomness a vital factor in building up the sampling distribution of means is the size of each of the r samples taken. Each sample must be of the same size and I have somewhat arbitrarily chosen a sample size of 30 as this is often seen to be the approximate dividing line between a small sample and a large one. If the population being sampled is itself near normal in shape the sampling distribution of means approaches normality quicker than for other types of population distribution and in this case we can also use a sample size of less than 30 and still achieve normality relatively quickly.

So, the process of repeated random sampling from a large population can give rise to a new probability distribution called the sampling distribution of means. Here we should note that we could, in the same way, have found other characteristics of the samples, say their median, mode or standard deviation, and these would have given rise to other sampling distributions, such as the sampling distribution of medians or the sampling distribution of standard deviations.

The central limit theorem tells us not only that the sampling distribution of means approaches normality as r, the number of samples taken, becomes large but that it is related to the parent population from which the samples were drawn as follows.

1 The mean of the sampling distribution of means, $\bar{X}$, is equal to the mean of the population, μ. That is, $\mu = \bar{X}$.
2 The standard deviation of the sampling distribution of means, S, is related to the population standard deviation, σ, by the following formula:

$$S = \frac{\sigma}{\sqrt{N}} = \frac{\text{population standard deviation}}{\text{square root of the number in the sample}}$$

The standard deviation S of the sampling distribution of means, is sometimes known as the standard error.

Let us now use Minitab to demonstrate these relationships between the sample statistics and the population parameters. We will build up two sampling distributions of means by taking repeated random samples from two types of population, first a non-normal population, a uniform distribution, and second a normal distribution.

A sampling distribution of means from a uniform distribution $U(a, b)$ I have chosen to use a uniform distribution as it is very far from normal in shape, and yet we will find that a sampling distribution of means drawn from it has a good degree of normality. Its approach to normality can be checked using NSCORES or MPLOTS of suitable normal parameters, but I have not done this here.

The uniform distribution $U(a, b)$ is a family of distributions which when plotted only take on values between $X = a$ and $X = b$. In other places on the variable X axis the distribution does not exist. It has the characteristic block or flat-top shape shown below in figure 9.8, and is as distinctive, in its own way, as the normal curve.

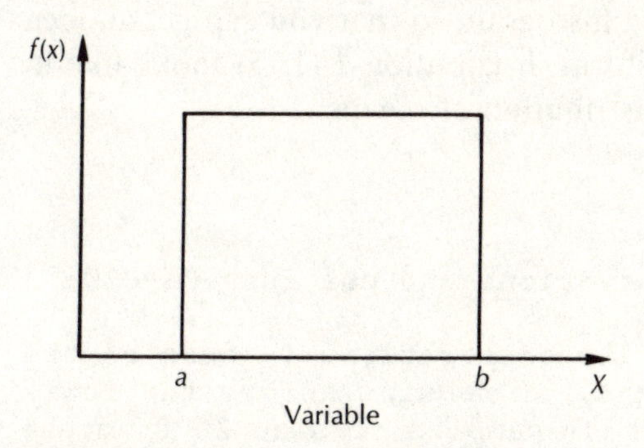

Figure 9.8 Uniform distribution U(a, b)

From the family of uniform distributions $U(a, b)$ I have chosen $U(1, 5)$, i.e. the values of the X variable at a and b are 1 and 5 respectively. The distribution or PDF stretches for 4 units along the X axis and, as the total area enclosed by any PDF must be one square unit, it is 1/4 unit in height.

In this example I have taken a number of random samples from the above uniform distribution, found the mean of each of these samples and then plotted these means. As you will see the distribution of these sample means has a shape that is close to the normal, despite the fact that the random sample was taken from the very distinctively shaped uniform distribution $U(1, 5)$.

Let me say a few words about some of the commands used, and then I would like you to follow the program through on the computer. The command RANDOM in Minitab can be used with many types of distribution in order to generate random observations from these distributions. We will generate 100 random values from our chosen Uniform distribution $U(1, 5)$ in each of the first 20 columns of the editor. We will then read these random values across, i.e. as rows, to get 100 rows or samples each of 20 values. After this we will find the mean of each of these rows or samples, put them in another column and then finally plot them. The plot will show the sampling distribution of means.

The command BASE 1000 tells the random number generator to start at a certain place and if you wish your working to be the same as mine, i.e. the same 'random' samples, you should use it.

In addition to plotting the sampling distribution of means I have then stacked all the 20 columns each containing 100 observations with the STACK command and put them in column C22. I have then illustrated C22 using a histogram so that you can get an idea of what the simulated uniform distribution $U(1, 5)$ looks like in relation to the sampling distribution of means.

```
MTB > BASE 1000
MTB > RANDOM 100 observations and put into C1-C20;
SUBC> UNIFORM A=1 B=5.
MTB > #This gives us 100 observations in each of the
MTB > #first 20 columns, or reading across the rows
MTB > #we have 100 random samples of size 20 from the
MTB > #uniform distribution U(1,5).
MTB > RMEAN C1-C20 and put into C21
MTB > #This finds the mean of each sample of 20 (i.e.
MTB > #each row) and puts it in C21.
MTB > HISTOGRAM C21
```

```
Histogram of C21 N = 100

Midpoint   Count
    2.2        1 *
    2.4        2 **
    2.6        7 *******
    2.8       27 ***************************
    3.0       30 ******************************
    3.2       23 ***********************
    3.4        7 *******
    3.6        2 **
    3.8        1 *

MTB > #This distribution looks reasonably normal
MTB > MEAN C21
   MEAN    =     2.9879
MTB > STDEV C21
   ST.DEV. =     0.25934
MTB > STACK C1-C20 and put into C22
MTB > HISTOGRAM C22

Histogram of C22 N = 2000
Each * represents 5 obs.

Midpoint Count
    1.2   213 *******************************************
    1.6   209 ******************************************
    2.0   177 ************************************
    2.4   224 *********************************************
    2.8   193 ***************************************
    3.2   202 *****************************************
    3.6   176 ************************************
    4.0   200 ****************************************
    4.4   213 *******************************************
    4.8   193 ***************************************

MTB > #As we see the uniform distribution has a more
MTB > #or less flat-top profile, whilst the sampling
MTB > #distribution drawn from it is near normal.
MTB > MEAN C22
   MEAN    =     2.9879
MTB > STDEV C22
   ST.DEV. =     1.1576
```

As can be seen from the histogram of the sampling distribution of means, random sampling can give a reasonable approximation to the normal distribution even when we start off with something like a uniform distribution as our parent population. Let us now take similar samples from a parent population which is itself normal and see what the resulting sampling distribution of means looks like.

A sampling distribution of means from a normal distribution, N(60, 3) I have used a similar program to one we used above for the uniform distribution but this time using a normal distribution with a mean of 60 and a standard deviation of 3 as the parent population.

You should read through the program and then try it for yourself. Don't forget to clear the columns of the data editor before starting. Again if you wish to replicate my samples you should also start the random number generator at the same point using the command BASE 1000.

```
MTB > ERASE C1-C22
MTB > BASE 1000
MTB > RANDOM 100 observations and put into C1-C20;
SUBC> NORMAL MU=60 SIGMA=3.
MTB > RMEAN C1-C20 and put into C21
MTB > #As with the previous distribution this
MTB > #finds the mean of each of the 100 samples
MTB > #of 20 that we have in each row and puts
MTB > #this mean in C21, i.e. C21 contains the
MTB > #sampling distribution of means for this
MTB > #normal distribution.
MTB > HISTOGRAM C21

Histogram of C21 N = 100

Midpoint  Count
    58.0      2 **
    58.4      3 ***
    58.8      6 ******
    59.2     16 ****************
    59.6     10 **********
    60.0     31 *******************************
    60.4      9 *********
    60.8     14 **************
    61.2      6 ******
    61.6      3 ***
```

```
MTB > MEAN C21
   MEAN        =      59.943
MTB > STDEV C21
   ST.DEV. =     0.79331
MTB > STACK C1-C20 and put into C22
MTB > HISTOGRAM C22
Histogram of C22 N = 2000
Each * represents 10 obs.
Midpoint Count
      48     1 *
      50     2 *
      52    20 **
      54    90 *********
      56   221 ***********************
      58   426 *******************************************
      60   485 *************************************************
      62   432 ********************************************
      64   232 ************************
      66    69 *******
      68    19 **
      70     2 *
      72     1 *

MTB > MEAN C22
   MEAN        =      59.943
MTB > STDEV C22
   ST.DEV. =       3.0901
```

Again the sampling distribution of means in C21 gives, when plotted, a reasonably good approximation to a normal distribution; however, it is interesting to note that by chance it is not as normal in form as the sampling distribution of means that we earlier derived from the uniform distribution by random sampling. As I mentioned previously these two distributions can be checked for normality using NSCORES or MPLOT techniques.

Let us now look at the relationship between the population parameters and the sample statistics for the above probability distributions. The central limit theorem quoted earlier predicted the following relationships: mean $\bar{X}$ of the sampling distribution of means is equal to the population mean μ; and standard deviation S of the sampling distribution of means is equal to $\sigma/\sqrt{N}$ where σ is the standard deviation of the population and N is the sample size. Are these relationships approximately true for our two examples? From the computer print-out we have the following (note that in the case of the uniform distribution I have calculated σ from the 2000 random values stacked in C22).

Uniform distribution $U(1, 5)$	*Normal distribution* $N(60, 3)$
Population	Population
$\mu = 3.00$	$\mu = 60$
$\sigma = 1.1576$	$\sigma = 3$
Sampling distribution of means	Sampling distribution of means
$\bar{X} = 2.9879$	$\bar{X} = 59.943$
$S = 0.25934$	$S = 0.79331$

As can be seen the means of the parent populations and of the respective sampling distributions of means are very similar in value. Let us now look at the standard deviations.

For the uniform distribution we have

$$\frac{\sigma}{\sqrt{N}} = \frac{1.1576}{\sqrt{(20)}} = 0.2588$$

which is close to the value of 0.25934 that we obtained. For the Normal distribution we have

$$\frac{\sigma}{\sqrt{N}} = \frac{3}{\sqrt{(20)}} = 0.6708$$

which again is relatively close to the value of 0.7933 that we obtained (bearing in mind that we only had 20 values in each sample and samples of less than 30 are usually considered to be small).

Thus these computer simulations provide us with a good illustration of the validity of the central limit theorem. By taking larger samples or by increasing the number of samples in our sampling distributions of means we could have made even closer inferences regarding the parent populations. With larger numbers we could also have obtained sampling distributions which gave even closer approximations to normality.

Single samples We now turn from repeated random sampling which can be used, as we have seen, to build up probability distributions known as sampling distributions, and we look at the more usual case where we only have a single sample.

Let us imagine that we take a single sample of size 20 ($N = 20$) in a random way from a large population. We can find its mean and call it $\bar{X}_1$ as before. We will also find the standard deviation of the sample and call it S_1. What can such sample statistics, $\bar{X}_1$ and S_1, tell us about the corresponding population parameters μ and σ? Remember that we now only have one single sample of 20 whereas in our previous work we had 100 samples each of size 20 and a derived probability distribution called the sampling distribution of means.

However, despite our single sample we do know, from the central limit theorem and indeed from our earlier simulations, that if we had a large number of such samples and calculated and plotted their means they would form a near-normal distribution with a mean of approximately μ (the population mean) and a standard deviation S, approximately equal to $\sigma/\sqrt{N}$ (the population standard deviation divided by the square root of the number in the sample). We only have one such sample but we can say that it lies somewhere within the normally distributed sampling distribution of means. Where within this distribution is it likely to lie? We cannot tell for certain, but the probability is that it will lie in middle or main body of the distribution rather than in one of the thin tails.

Confidence Intervals for Means

From our knowledge of the normal distribution we can make certain probability predictions. We can say that our single random sample value will lie within approximately 2 standard deviations of the population mean μ, in 95% of such samples. That is, there is only about a 5% chance of our single sample lying in either of the tails of the sampling distribution of means which we can imagine it to be a part of. The population mean μ is likely to lie within approximately 2 standard deviations (the standard deviation of the sampling distribution of means) of our single sample mean with a probability of 0.95 (figure 9.9).

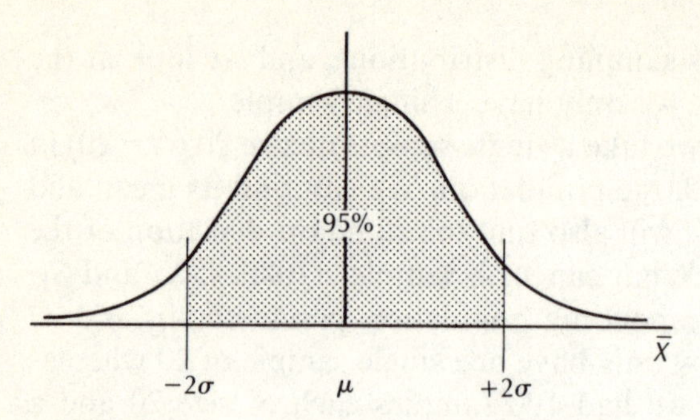

Figure 9.9 A sampling distribution of means

The statement we have made above is a probability statement or a statistical inference concerning the mean of the population based on our single sample and its mean. We could of course also frame other general probability statements about the population mean by considering not 95% of the normal distribution but say 68% or 99.7%. Also, as we shall see in the next chapter, with the help of statistical tables we can make statistical inferences at whatever level of probability that we choose.

To return to our example, we have said that the population mean μ that we are trying to estimate lies within 2 standard deviations either side of our single sample mean with a probability of 0.95. This type of probability statement or statistical inference is often called a confidence interval. Somewhat roughly but for all practical purposes in an introductory text we can say that we are confident, at a probability level of 0.95, that the population mean μ will lie in the given interval, window or corridor. We frequently use such constructions in everyday life when we say, for instance, that we will meet a friend at a certain place at 6 o'clock, give or take 10 minutes. That is, we specify a place and a 20 minute interval, window or corridor of time in which we will probably arrive.

Our estimate of the population mean μ is as follows:

$$\mu = \bar{X}_1 \pm 2\,S \qquad (P = 0.95)$$

$\bar{X}_1$ is the single sample mean, S is the standard deviation of the sampling distribution of means and the probability given at the end estimates that we expect to be correct in our estimate in about 95% of cases. We must also be prepared to be in error in about 5% of cases.

However, by the central limit theorem we know that the standard deviation S, of the sampling distribution of means is equal to $\sigma/\sqrt{N}$. When we make this substitution in the above expression we get

$$\mu = \bar{X}_1 \pm \frac{2\sigma}{\sqrt{N}} \qquad (P = 0.95)$$

This is a very neat statement of inference, but as you can see there is one big drawback. We do not know the value of the standard deviation of the population σ. However, for large samples of 30 or more ($N \geqslant 30$) it is usual to assume that the standard deviation S_1 of the single sample and the standard deviation of the population from which the sample was drawn are equal, i.e. $\sigma = S_1$. This approximation generally works satisfactorily in practice and it can be justified in theoretical terms but we shall not do so here. We can now finally write our statistical inference as follows:

$$\mu = \bar{X}_1 \pm \frac{2S_1}{\sqrt{N}} \qquad (P = 0.95)$$

where, μ is the population mean, $\bar{X}_1$ is the single sample mean, S_1 is the single sample standard deviation and N is the number of observations in the single sample.

It is important to remember that such inferences or confidence intervals are probability statements. They are based on probabilities or bets on the correctness of our inference. We cannot expect to be always correct in our prediction. We may in some cases be wrong. However, in the majority of cases, given that our assumptions are correct and that we have done our sums properly, we expect to be right in our predictions. In the above case we expect to be correct in our prediction in the long run in about 95 cases in every 100 ($P = 0.95$). We must also be prepared to be wrong in about 5 cases in every 100.

Before looking at an example using the above formula it is worth noting one fact about the standard deviation of the sample. It has been found that the best results, i.e. the most accurate predictions, are obtained by slightly modifying the basic standard deviation formula by using a denominator of $N-1$ rather than N. That is, the modified formula is

$$\text{sample standard deviation} = S_1 = \sqrt{\left[\frac{\sum(X - \bar{X})^2}{N-1}\right]}$$

On most calculators you will find that this sample standard deviation is usually labelled $\sigma_{n\text{-}1}$.

Let us end this chapter with a simple worked example of the use of confidence intervals.

Example A random sample of 32 heights of female undergraduates in the South Side University was obtained. The mean of the sample was found to be 65 inches and its standard deviation 3 inches. Use this single random sample to estimate the height of female undergraduates in this University at the 0.95 level of probability.

$$\mu = \bar{X}_1 \pm \frac{2S_1}{\sqrt{N}} = 65 \pm \frac{2 \times 3}{\sqrt{(32)}} = 65 \pm 1.1 = 63.9 \text{ to } 66.1 \text{ inches}$$

Our confidence interval for the population mean height at this level of probability ($P = 0.95$) is 63.9 to 66.1 inches.

Postscript

Our focus in this chapter has been the normal distribution. This probability distribution is of fundamental importance in sampling and in the statistics of inference. We looked at various ways in which the distribution can arise or be used in data analysis. We found very good approximations to this distribution in the natural world when we examined adult heights and we also showed by computer simulation that the binomial distribution approaches the normal as the number of trials increases.

Finally we examined how the normal distribution can arise from random sampling and looked at the relationship between sample and population using the intermediate concept of the sampling distribution. We ended the chapter with the calculation of confidence intervals for the population mean based on a single sample.

We continue our study of the normal distribution in the next chapter by focusing on the standardized form of the distribution. As we shall find all normal and near normal probability distributions can easily be reduced to the standard normal form, enabling us to use one set of tables to relate Z scores, areas and probabilities for all such distributions.

Exercises

1 Ms XX and Mr XY decide to have a family of four children and hope to have at least three girls. Assuming that the probability of male and female births is equal, draw a probability tree and find the probability that they will have (i) three girls, (ii) four girls, (iii) at least three girls. Plot the probabilities of getting a girl in the form of a histogram.

2 Conduct a small experiment as follows to compare theoretical and empirical probability. Toss a single fair coin and note the number of heads that appear in every sequence of 6 tosses. Make 64 test runs, each of 6 tosses, tally the number of heads and fill in table 9.1.

Table 9.1

No.of heads	*Tally*	*Observed frequency*	*Observed relative frequency* $P_o(X)$	*Theoretical relative frequency* $P_t(X)$
0			/64	1/64
1			/64	6/64
2			/64	15/64
3			/64	20/64
4			/64	15/64
5			/64	6/64
6			/64	1/64

Obtain the observed frequency from the tally of heads and so find the observed relative frequency for each value of the variable (number of heads). I have given the relative theoretical frequencies that are to be expected from the binomial series in the last column. Draw both the theoretical and the observed probability distributions as histograms on the same axis.

3 Suppose that there is an even chance that a girl will be upwardly socially mobile relative to her father as measured by the Registrar General's system of social class (intergenerational mobility). In a randomly chosen family of three girls what is the theoretical probability that two will experience upward mobility?

4 Use Minitab to produce a binomial probability distribution for $N = 30$ and $P = 0.5$ (i.e. the pattern for 30 tosses of a fair coin). Find the parameters of a suitable normal distribution that could be used to model the binomial and plot both distributions on the same axis using the MPLOT command. Comment briefly on the goodness of fit of the model.

5 Using Minitab simulate a sampling distribution of means when 100 samples each of size 20 are taken randomly from a normal population which has the parameters $\mu = 68$ and $\sigma = 3.7$ (use BASE 1000). Find the mean and standard deviation of the sampling distribution of means and compare them with the mean and standard deviation of the parent population. Plot histograms of the sampling distribution of means and the parent population using a stack of all the 2000 random observations to give a simulated parent population.

6 The mean height and standard deviation of a random sample of the heights of 100 male students at the XYZ University were found to be 177 cm and 8 cm respectively. Calculate the 95% confidence limits for the mean of the population from which the sample was drawn.

7 A continuous random variable is normally distributed with a mean of 10 units and a standard deviation of 2 units. What is the probability that a randomly chosen value of the variable is greater than 12 units?

8 A few years ago the mean number of children per family in Britain was 2.6. At this time a random sample of 100 army officers was found to have a mean of 3.5 children with a standard deviation of 4. Give an estimate of the mean number of children in all army officer families at this time and on the basis of this estimate decide whether army officers have larger families than most. Use the 95% level for your estimate.

Answers to Exercises

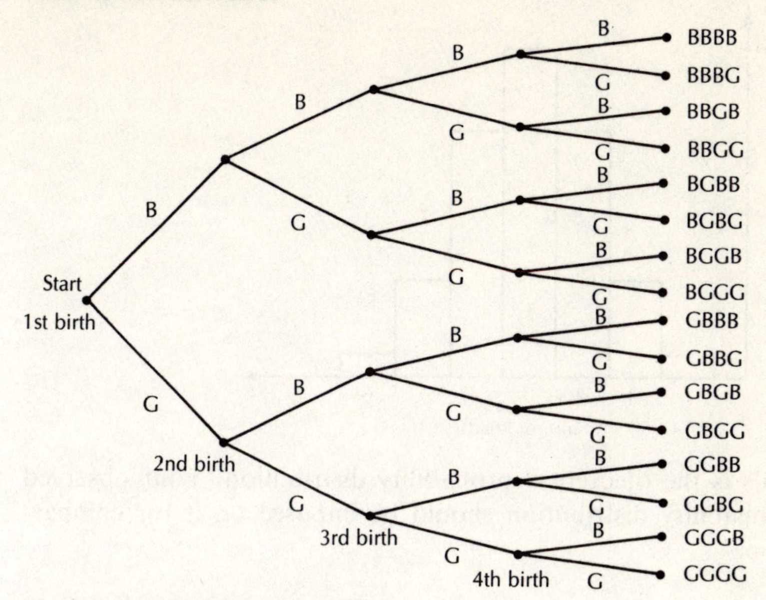

1 (i) P (3 girls) = 4/16 = 1/4
(ii) P (4 girls) = 1/16
(iii) P (at least 3 girls) = 4/16 + 1/16 = 5/16

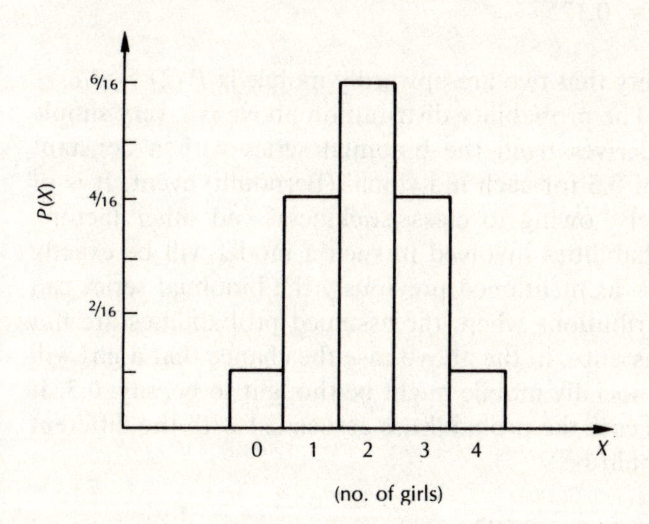

2

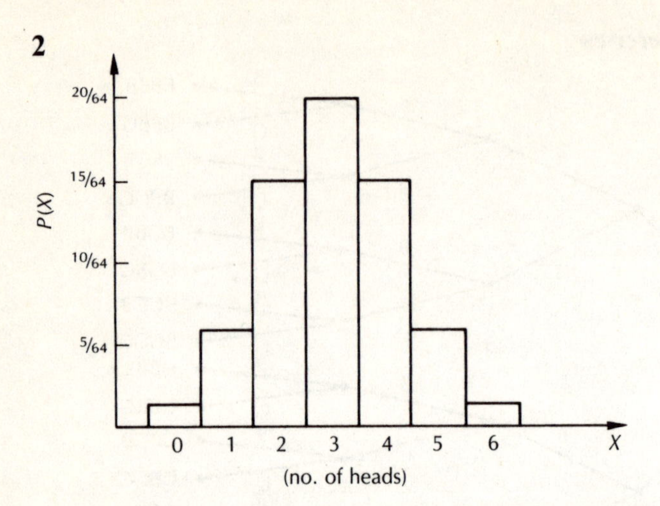

This is the theoretical probability distribution. Your observed probability distribution should be imposed on it for comparison.

3 $P(0) = 1/8 = 0.125$
$P(1) = 3/8 = 0.375$
$P(2) = 3/8 = 0.375$
$P(3) = 1/8 = 0.125$

The probability that two are upwardly mobile is $P(2) = 3/8 = 0.375$. *Note*: The probability distribution above is a very simple model and derives from the binomial series with a constant probability of 0.5 for each individual (Bernoulli) event. It is of course unlikely, owing to class 'stickiness' and other factors, that the probabilities involved in such a model will be exactly 0.5. However, as mentioned previously the binomial series can generate distributions where the assumed probabilities are not equal. For instance, in the above case the chance that a girl will be upwardly socially mobile might be thought to be, say, 0.3. If this were the case the probabilities associated with the different outcomes would be

$$P(0) = 1(.7)^3 = 0.343$$
$$P(1) = 3(.3)^1(.7)^2 = 0.441$$
$$P(2) = 3(.3)^2(.7) = 0.189$$
$$P(3) = 1(.3)^3 = 0.027$$

4 Your 'program' for this question should look something like this.

$$MU = NP = 30 \times 0.5 = 15$$

$$SIGMA = \sqrt{[NP(1-P)]} = \sqrt{[(30)\times 0.5 \times 0.5]} = \sqrt{7.5} = 2.74$$

```
MTB > SET C1
DATA> 0:30
DATA> END
MTB > PDF C1 and put into C2;
SUBC> BINOMIAL N=30 P=.5.
MTB > #With our calculated values of MU(15)
MTB > #and SIGMA(2.74) we can now fit a suitable
MTB > #normal distribution to the binomial.
MTB > PDF C1 and put into C3;
SUBC> NORMAL MU=15 SIGMA=2.74.
MTB > MPLOT C2 against C1 and C3 against C1
```

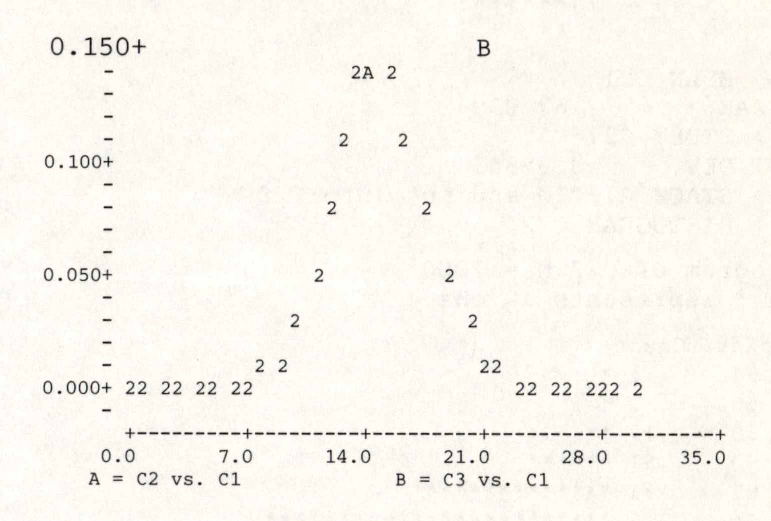

As can be seen the normal distribution N(15, 2.74) provides a good fit to the binomial B(30, 0.5).

5

```
MTB > BASE 1000
MTB > RANDOM 100 observations and put into C1-C20;
SUBC> NORMAL MU=68 SIGMA=3.7.
MTB > RMEAN C1-C20 and put into C21
MTB > #The above treats each row as a sample of
MTB > #20, finds the mean of each of these samples
MTB > #and puts it into C21. C21 is therefore the
MTB > #sampling distribution of means.
MTB > HISTOGRAM C21

Histogram of C21 N = 100
Midpoint  Count
    65.5      2 **
    66.0      3 ***
    66.5      6 ******
    67.0     14 **************
    67.5     12 ************
    68.0     31 *******************************
    68.5     10 **********
    69.0     13 *************
    69.5      7 *******
    70.0      2 **

MTB > MEAN C21
   MEAN    =    67.930
MTB > STDEV C21
   ST.DEV. =   0.97841
MTB > STACK C1-C20 and put into C22
MTB > HISTOGRAM C22

Histogram of C22 N = 2000
Each * represents 10 obs.

Midpoint  Count
      54      1 *
      56      3 *
      58     17 **
      60     51 ******
      62    131 **************
      64    235 ************************
      66    369 *************************************
      68    397 ****************************************
      70    384 ***************************************
      72    229 ***********************
      74    123 *************
      76     41 *****
      78     16 **
      80      2 *
      82      0
      84      1 *
```

The mean and standard deviation of the sampling distribution of means are 67.930 and 0.97841 respectively. This mean is relatively close to the population mean of 68 and $\sigma/\sqrt{N} = 3.7/\sqrt{(20)} = 0.83$, which is reasonably close to our value of 0.98 bearing in mind that we used relatively small samples of size 20. If your computer has sufficient capacity you may care to repeat this exercise using samples of 30 or more to see what difference sample size makes to accuracy.

6 Population mean $= 177 \pm 2 \times 8/\sqrt{(100)} = 177 \pm 1.6$ $(P = 0.95)$. Confidence limits are 175.4 cm to 178.6 cm.

7 Mean, 10 units; standard deviation, 2 units. We know that for a normal distribution approximately 68% of the area lies within one standard deviation of the mean. Either side of this central area we must have two tails each of 16% of the area (68 + 16 + 16 = 100). The values of the variable that we are interested in must lie in the shaded right-hand tail to the right of the point where the standard deviation is +1.

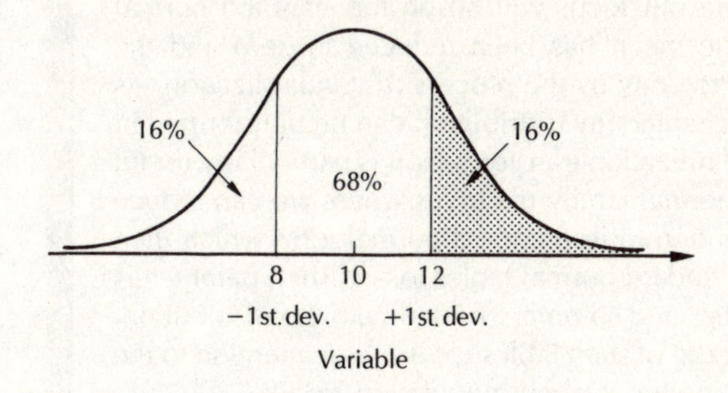

8 Population mean (army officers)

$$= 3.5 \pm 2 \times \frac{4}{\sqrt{(100)}} \text{ children.} \qquad (P = 0.95)$$
$$= 3.5 \pm 0.8 \text{ children.}$$
$$= 2.7 \text{ to } 4.3 \text{ children.}$$

We are told that the general population mean is 2.6 and this falls outside the confidence limits that we have calculated for army officers' families. It therefore appears at this level $(P = 0.95)$ that army officers do tend to have larger families.

10 Using Standard Normal Tables

In this chapter we continue our examination of the normal distribution but here our focus will be on the standard normal form, i.e. where the mean has been reduced to zero and the standard deviation to unity by the process of standardization. As we saw in the last chapter any distribution can be transformed in this way. The standardization transformation is particularly useful in the case of the normal family of curves where we can reduce any Normal distribution to its Standard normal form which then enables us to use standard normal tables to find such parameters as areas, probabilities and so on associated with the distribution. After looking at the use of such tables we turn our attention to the testing of simple hypotheses about population means.

In the last chapter we saw that the mathematical form of the normal distribution can be used to model many near-normal distributions that arise in various ways. We noted that each member of the family of normal curves can be defined uniquely in terms of two parameters, the population mean μ and the standard deviation σ. Each normal or near-normal distribution can be standardized and reduced to the useful standard normal form by the transformation

$$X \rightarrow \frac{X-\mu}{\sigma}$$

This transform gives us the Z score, i.e.

$$Z \text{ score} = \frac{X-\mu}{\sigma}$$

In chapter 9 we looked briefly at the shape of this standard normal curve and at how the area and corresponding probabilities were distributed about the mean. We noted the symmetry of the distribution and the fact that virtually all (99.7%) of the area is contained within 3 standard deviations either side of the mean (Z score $+3$ to -3), while 95% of the area and probability are to be found with 2 standard deviations of the mean (Z score approximately $+2$ to -2).

As we also noted, the area enclosed by a probability distribution is one square unit and this can represent the sum of all possible probabilities (figure 10.1).

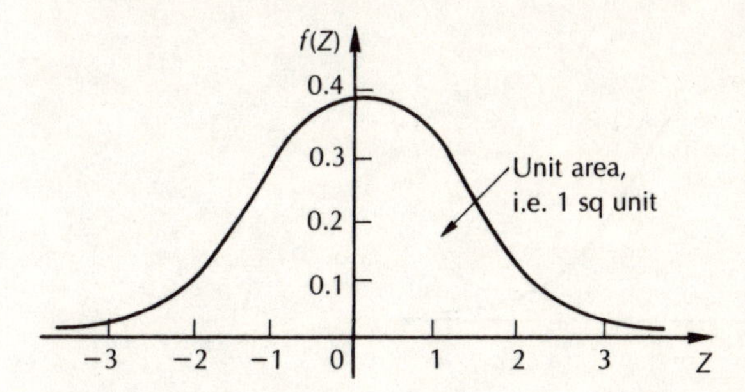

Figure 10.1 Probability density function, standard normal distribution N(0,1)

Areas and Probabilities

The area under the standard normal curve can be related to the Z scores. Each value of Z can be thought of as cutting off an area of the distribution that corresponds to a probability. Let us now consider a few specific cases based on what we already know about the normal distribution. In each case you should focus on the Z scores in the shaded areas.

1 As we saw in chapter 9, the area between $Z = -1$ and $Z = 1$ represents 68% of the area under the curve and so 68% of the total probability (figure 10.2).

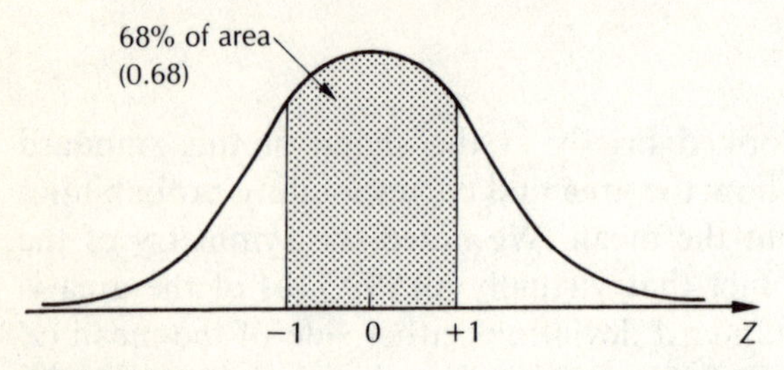

Figure 10.2

That is, $P(-1 \leqslant Z \leqslant 1) = 0.68$. The probability that the Z score lies between ± 1 is 0.68. (The symbol $\leqslant$ means less than or equal to.)

2 In figure 10.3 the shaded area occupies half the area under the curve.

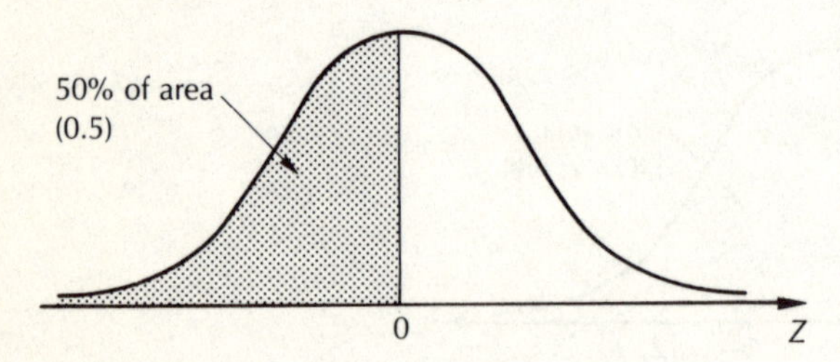

Figure 10.3

From the symmetry of the distribution we can say that $P(Z \leqslant 0) = 0.5$. That is, the probability that Z is less than or equal to zero is 0.5.

3 In figure 10.4 the shaded area in each tail of the distribution is equal to 2.5% of the area under the curve. The total shaded area (both tails) is therefore equal to 5% of the area.

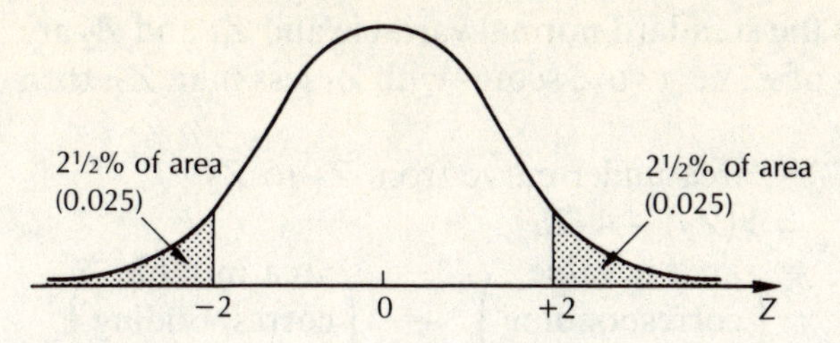

Figure 10.4

> The probabilities associated with the shaded tail areas can be expressed as follows: $P\ (Z \leqslant -2) = 0.025$ and $P\ (Z \geqslant 2) = 0.025$. (The symbol $\geqslant$ means greater than or equal to). That is, the probability that the Z score is less than or equal to -2 is 0.025 and the probability that the Z score is greater than or equal to 2 is 0.025.

The above probabilities arise from our previous knowledge of the characteristics of the standard normal distribution. Let us now use statistical tables to work out the areas/probabilities associated with any Z score. The table of probabilities that we use for the standard normal curve is to be found in Neave 1992 and in Table 1, appendix 2, p371–2. Neave labels the area corresponding to a Z score as $\Phi(Z)$ (pronounced phi zed). What is tabulated in Neave is the area less than or equal to a given Z score, i.e. $P(Z \leqslant Z_1) = \Phi(Z_1)$. The probability that the Z score is less than Z_1 is equal to the shaded area $\Phi(Z_1)$ (figure 10.5).

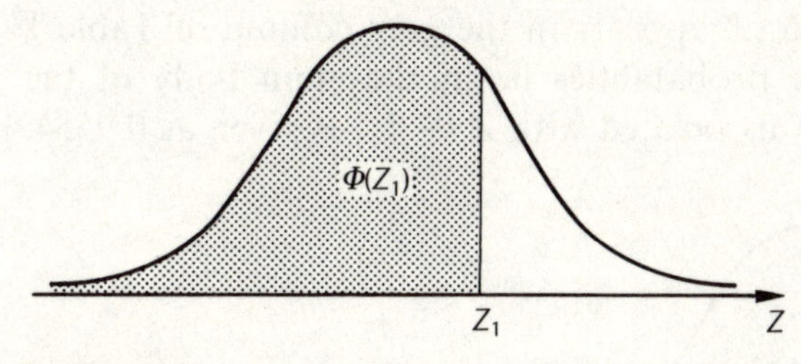

Figure 10.5

In general if Z is the standard normal variable and Z_1 and Z_2 are two specific values of Z, i.e. two Z scores with Z_1 less than Z_2, then

$$\begin{aligned} P(Z_1 \leqslant Z \leqslant Z_2) &= \text{area under curve from } Z_1 \text{ to } Z_2 \\ &= \Phi(Z_2) - \Phi(Z_1) \\ &= \begin{pmatrix}\text{area in table} \\ \text{corresponding} \\ \text{to } Z_2\end{pmatrix} - \begin{pmatrix}\text{area in table} \\ \text{corresponding} \\ \text{to } Z_1\end{pmatrix} \end{aligned}$$

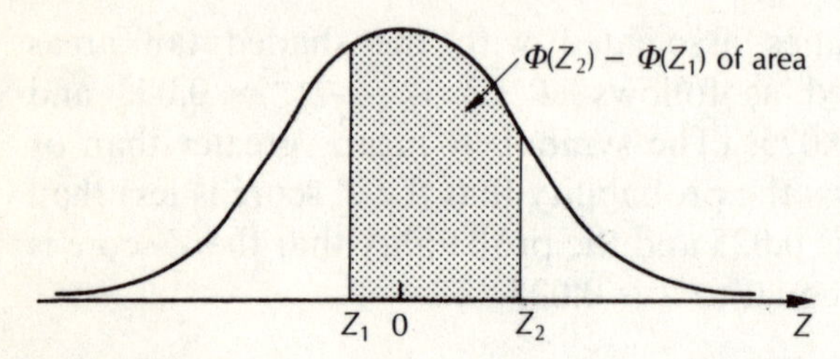

Figure 10.6

In other words, if we wish to find the area under the standard normal curve contained between two Z scores (shaded in figure 10.6) we find the $\Phi(Z)$ for each and take the difference. This shaded area corresponds to the probability that the Z score lies between Z_1 and Z_2. A few examples should make this clear.

Example 1 What is the probability that a Z score will lie between $Z = -0.4$ and $Z = 1.7$?

Looking up the probability associated with $Z = -0.4$ in Neave we find it to be 0.3446 (Z appears in the grey column of Table 1 Appendix 2 while the probabilities lie in the main body of the table). The probability associated with $Z = 1.7$ is given as 0.9554.

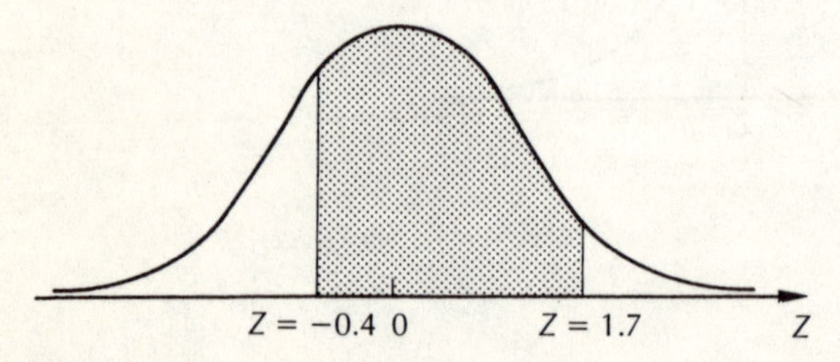

Figure 10.7

Taking the difference between these two Φ values we get the probability that the Z score lies in the shaded area between $Z = -0.4$ and $Z = 1.7$ (figure 10.7).

$$\begin{aligned} P(-0.4 \leqslant Z \leqslant 1.7) &= \Phi(1.7) - \Phi(-0.4) \\ &= 0.9554 - 0.3446 \\ &= 0.6108 \end{aligned}$$

The probability that Z lies between -0.4 and 1.7 is 0.6108, or approximately 61%.

Example 2 What is the probability that the Z score will lie between $Z = 0.6$ and $Z = 2.5$?

Looking up the probabilities associated with the two Z scores we find them to be 0.72570 and 0.99379. The difference between these two figures gives us the probability associated with the shaded area (figure 10.8):

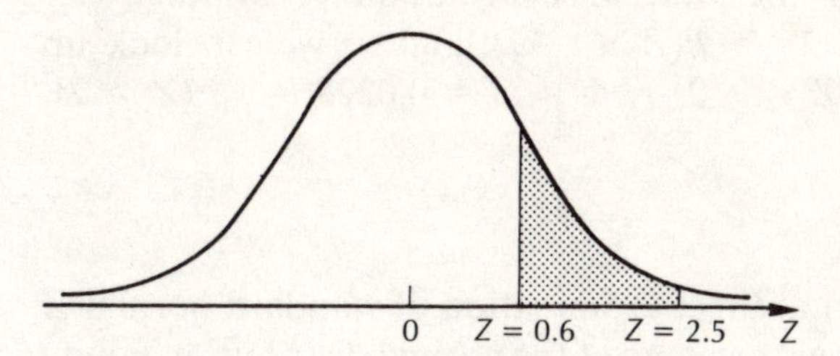

Figure 10.8

$$\begin{aligned} P\,(0.6 \leqslant Z \leqslant 2.5) &= \Phi(2.5) - \Phi(0.6) \\ &= 0.99379 - 0.72570 \\ &= 0.2681 \end{aligned}$$

The probability that Z lies in the shaded area is 0.2681, or approximately 27%.

Example 3 Let us use the tables to find another probability that we looked at in the last chapter. What is the probability that Z is greater than 2?

Here we require the probability that Z is greater than 2, i.e. $P(Z > 2)$, but what Neave tabulates is the 'less than' value so we need to subtract this from the total probability of one to get the probability that Z lies in the shaded area (figure 10.9), i.e.

$$P(Z > 2) = 1 - P(Z \leqslant 2) = 1 - \Phi(2) = 1 - 0.9772 = 0.0228,$$

or approximately 23%.

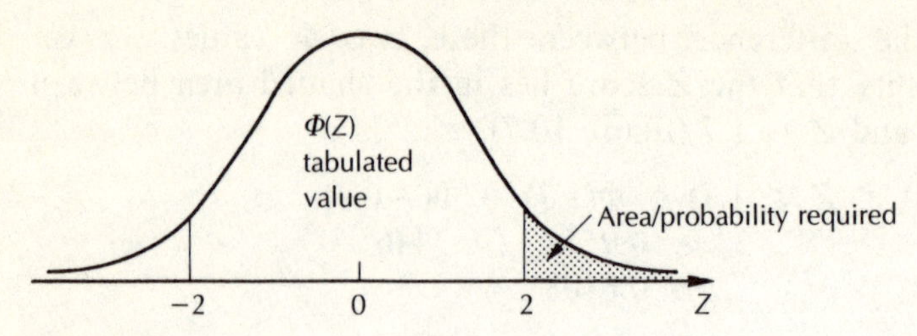

Figure 10.9

This is of course close to the approximate probability 0.025 (2.5%) which we found earlier. (We could note in passing that a Z score of 1.96 rather than 2 would give us a tail area of exactly 2.5%.)

The logic of the process in this example is clear, but it is probably easier to note that as the distribution is symmetrical the two tails cut off by Z and $-Z$ are equal in area and therefore represent the same probability as the normal distribution is symmetrical. For example $P(Z > 2) = P(Z \leqslant -2)$, and so we can look up $P(Z \leqslant -2)$ to get $P(Z \leqslant -2) = \Phi(-2) = 0.0228 = P(Z > 2)$.

Transformation

Let us now turn from a direct examination of standard normal Z scores and look at other members of the normal distribution family and how we can use the tables to evaluate parts of their areas and corresponding probabilities. Provided that we know the parameters of these normal distributions, i.e. the mean μ and the standard deviation σ, we can easily transform them into the standard normal form. As previously the appropriate transform is

$$X \rightarrow \frac{X-\mu}{\sigma}$$

Let us look at an example of such a transformation.

Example 4 A large population of IQ scores are distributed in a near-normal fashion with a mean of 100 and a standard deviation of 12. What proportion of the population are likely to have

(a) IQ scores between 90 and 115?
(b) IQ scores greater than 115?

If we assume that the original distribution can be modelled by a normal distribution with mean $\mu = 100$ and standard deviation $\sigma = 12$, we can transform this to the standard normal form and so work out the relevant areas and probabilities.

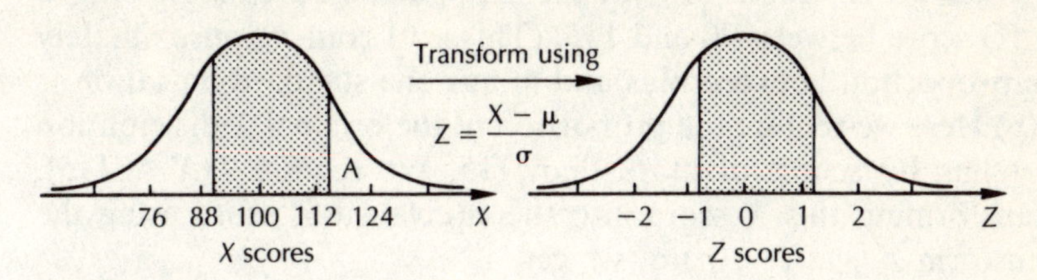

Figure 10.10

(a) We wish to know the proportion of the original X distribution cut off by the X scores of 90 and 115, i.e. the shaded area in figure 10.10. What we need is the probability $P(90 \leqslant X \leqslant 115)$. We cannot find this directly but we can standardize by transforming each part of the inequality with our formula $Z = (X - \mu)/\sigma$.

As we treat each part of the inequality in the same (positive) way we do not alter its validity and each X score now becomes transformed into a Z score. We can now, from Neave, find the probabilities associated with each of these Z scores.

$$P(90 \leqslant X \leqslant 115) = P\left(\frac{90 - 100}{12} \leqslant \frac{X - 100}{12} \leqslant \frac{115 - 100}{12}\right)$$

$$= P\left(\frac{-10}{12} \leqslant \frac{X - \mu}{\sigma} \leqslant \frac{15}{12}\right)$$

$$= P(-0.833 \leqslant Z \leqslant 1.25)$$

We now have a probability statement involving the variable Z rather than X so we can find the probabilities from Neave as before:

$$\begin{aligned} P(-0.833 \leqslant Z \leqslant 1.25) &= \Phi(1.25) - \Phi(-0.833) \\ &= 0.8944 - 0.2025 = 0.6919 \end{aligned}$$

That is

$$P(90 \leqslant X \leqslant 115) = P(-0.833 \leqslant Z \leqslant 1.25) = 0.6919$$

In other words, about 69% of the X population are likely to have an IQ score between 90 and 115. (This is of course approximately the proportion between plus and minus one standard deviation.)

(b) Here we require the proportion of the original X distribution that has IQ scores of more than 115. We require $P(X > 115)$. Transforming this X score into the equivalent Z score using the transform $Z = (X - \mu)/\sigma$, we get

$$P(X > 115) = P\left(\frac{X - 100}{12} > \frac{115 - 100}{12}\right)$$

As before, treating each part of the inequality in the same positive manner does not affect its validity, and so we get

$$P(X > 115) = P\left(Z > \frac{15}{12}\right) = P(Z > 1.25)$$

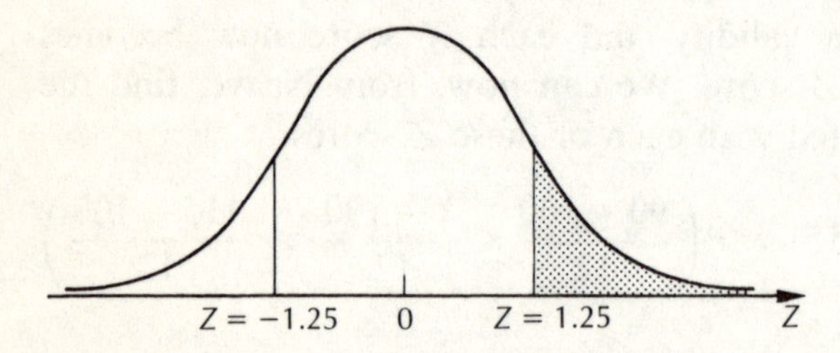

Figure 10.11

This probability corresponds to the area of the distribution to the right of $Z = 1.25$ (figure 10.11). We know by symmetry that this is equal to the area to the left of $Z = -1.25$, i.e.

$$P(X > 115) = P(Z > 1.25) = P(Z \leqslant -1.25)$$

$$= \Phi(-1.25) = 0.1056$$

That is, about 10.5% of the X population have an IQ greater than 115.

Hypothesis Testing

We now turn our attention to testing hypotheses. As this is an introductory text we confine ourselves to testing hypotheses about population means. The process of testing such hypotheses, as you will see, is similar and cognate with the way we found confidence intervals in the last chapter. Both are ways of expressing statistical inference. That is, they are both ways of moving from our firm knowledge of sample characteristics (sample statistics) to an understanding of the parameters of the population from which the sample was drawn.

We look at the case where a single large sample ($N \geqslant 30$) has been drawn randomly from the parent population. After finding the mean of this single sample (and its standard deviation) we treat it as a single element in the sampling distribution of means. As we saw in the last chapter the sampling distribution of means is a probability distribution which, when many sample means of the same size are taken, approaches a normal distribution.

We can think of this sampling distribution of means as being the intermediate link between our knowledge of the single sample and our predictions about the population. It is this intermediary probability distribution which bears the weight of the statistical inference.

We know that our single sample mean lies theoretically within the sampling distribution of means. Where exactly within this probability distribution it lies we cannot tell; however, from our knowledge of the normal shape of the sampling distribution of means we can estimate its probable distance from the mean.

The three sets of means and standard deviations are labelled as follows.

Single sample		*Sampling distribution of Means*		*Parent Population*
Mean $\bar{X}_1$	→	Mean $\bar{X}$	→	Mean μ
St. devn. S_1		St. devn. $S = \dfrac{\sigma}{\sqrt{N}}$		St. devn. σ
No. in sample N		No. in each sample N		

As previously we start by using the Z score. We call this our test statistic. By the central limit theorem and our previous approximations we have

$$Z = \frac{\bar{X} - \mu}{S} = \frac{\bar{X} - \mu}{\sigma/\sqrt{N}}$$

However, generally we do not know the population standard deviation and so we will also say that the Z score is approximately equal to the following:

$$Z = \frac{\bar{X}_1 - \mu}{S_1/\sqrt{N}}$$

where $\bar{X}_1$ is the single sample mean, S_1 is the single sample standard deviation, μ is the mean of the population and N is the number of observations in the single sample.

We calculate the test statistic or Z score based on the assumption that the null hypothesis H_0 is true and then compare it with a critical value of Z obtained from probability tables. We then make decision to either accept or reject the null hypothesis.

Let us illustrate the process by considering an example.

Example 5 Judo players are particularly prone to athlete's foot. Let us say that it has been found from long experience that the condition can be cleared up in ten days with the application of Tinglederm. The firm producing this product has developed another foot-rot cream, Supertinglederm, which they tested on 30 randomly selected male and female players. They found that the mean clear-up time for this new product is 9 days with a standard deviation of 2.4 days. Are they right, on the basis of this sample, to claim that Supertinglederm is more effective?

I will outline the steps we need to take to evaluate their claim.

1 State the hypothesis to be tested and its alternative The hypothesis H is usually stated in the null or no-difference form H_0. In this case we assume, for the hypothesis, that Supertinglederm is no more effective than Tinglederm and has the same mean clear-up time of 10 days. That is, we are going to assume that our random sample is drawn from this parent population with a mean of 10. As an alternative hypothesis H_1 we will say that the mean clear-up time is less than 10 days.

H_0: $\mu \geqslant 10$ days
H_1: $\mu < 10$ days

It is important to note that both the null and alternative hypotheses are always about *population* parameters. Here we are testing a hypothesis about the *population* mean μ.

The above alternative hypothesis leads to a 'one-tail test', and we employ a critical region in one tail of the test statistic distribution. That is, the alternative hypothesis has a 'single direction' implied by the words 'less than ' or 'lower'. Similarly if the alternative hypothesis had used words such as bigger, heavier, shorter or other words indicating movement in *one direction only* then we would generally use a one-tail test with the critical region in one tail of the distribution.

However, if we had used $\mu \neq 10$ (μ is not equal to 10) as an alternative hypothesis, then it would have been appropriate to use a two-tail test with critical regions in both tails of the test statistic distribution as no single direction is implied in the alternative hypothesis. If μ is not equal to 10 it could be either bigger or smaller and so a two-tail test is appropriate. However, in the case we are now considering one single direction, less than, is implied and so we will use a one-tail test.

2 State the test statistic to be used and its probability distribution As mentioned earlier our test statistic will be

$$\frac{\bar{X}_1 - \mu}{\bar{S}_1/\sqrt{N}} = Z$$

and this has a standard normal or Z distribution.

$\bar{X}_1$ sample mean
S_1 sample standard deviation
μ mean of the population if H_0 is true
N number in the sample

We should note here that the test statistic Z should only be applied to a hypothesis when the random sample is large ($N \geqslant 30$). If the size of the sample is less than 30 we cannot usually make the assumption, on which the test rests, that the sampling distribution of means is approximately normal in shape.

3 Choose the value of the significance level α, and determine the corresponding critical region The significance level α, that we will choose is 0.05 to give us $P = 0.95$ as we had in chapter 9. We can now draw a sketch with the shaded critical region or tail on the left (figure 10.12).

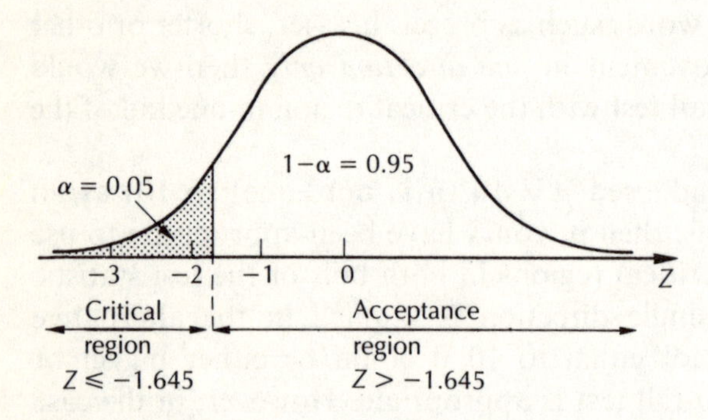

Figure 10.12

The Z score of $Z = -1.645$ forms a boundary to this critical area. Five per cent, i.e. 0.05, of the area of the distribution lies to the left of this Z value of -1.645. This Z value is found by reading the standard normal tables (Appendix 2, Table 1) in 'reverse'. We find the value of 0.05 in the body of the table and find that it corresponds to a Z score of -1.645. In this case the region to the left of -1.645 is known as the critical region and if our value of the test statistic Z falls within it we will reject the null hypothesis H_0, i.e. $\mu \geqslant 10$. We will then accept the alternative hypothesis H_1 that the mean of the population is less than 10 days.

We could of course have chosen a different and more rigorous significance level, say $\alpha = 0.01$ (1%) corresponding to a probability of $P = 0.99$. This would have given us a smaller critical region and so made it more difficult to reject H_0.

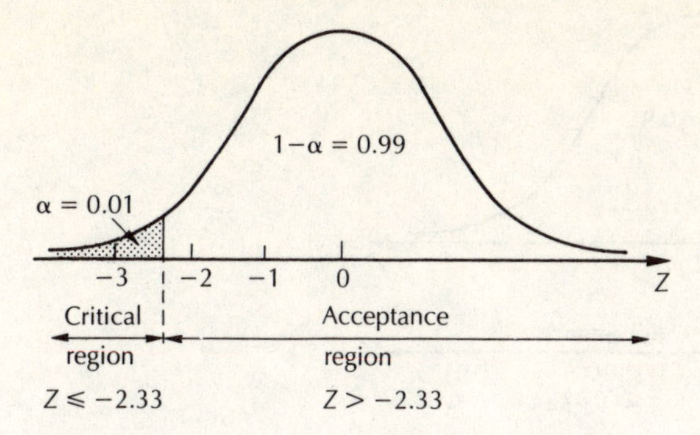

Figure 10.13

If we had chosen this more rigorous significance level our diagram and the corresponding critical region would have appeared as above in figure 10.13. The bounding Z score is now $Z = -2.33$. However, let us continue our example with the chosen 5% significance level, i.e. $\alpha = 0.05$.

4 State the decision rule for the acceptance or rejection of H_0

reject H_0 if $Z < -1.645$ $\qquad \alpha = 0.05$ $(P = 0.95)$
accept H_0 if $Z \geqslant -1.6456$

5 Calculate the value of the test statistic Z on the assumption that H_o is true

$$\text{test statistic } Z = \frac{\bar{X}_1 - \mu}{S_1/\sqrt{N}} = \frac{9 - 10}{2.4/\sqrt{30}} = -2.28$$

6 Make the decision to accept or reject H_0 In this case (figure 10.14) our test statistic $Z = -2.28$ does lie in the critical region for $\alpha = 0.05$ $(P = 0.95)$ and so we must reject the null hypothesis H_0 that $\mu \geqslant 10$. Our result is significant at this level. Our random sample is probably drawn from a population which has a mean of less than 10 days. Supertinglederm does appear, at this level, to have a significantly quicker clear-up time.

However, before turning from this example we should note that had we adopted the more rigorous significance level of $\alpha = 0.01$ $(P = 0.99)$ the limit of the critical region would have been

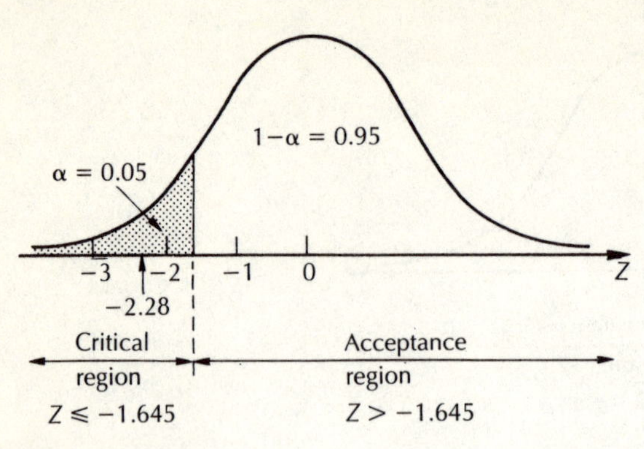

Figure 10.14

$Z= -2.33$. Our value of the test statistic, $Z= -2.28$, does not fall in the critical region and therefore the hypothesis H_0, $\mu \geqslant 10$, would still stand. There is insufficient evidence to reject it at this level of $\alpha = 0.01$ ($P = 0.99$) there is no strong evidence that it works and we can conclude that Supertinglederm does *not* appear to have a significantly quicker clear-up time.

The significance level we choose in step 3 depends on many factors. The $\alpha = 0.05$ is a very common level but if the decision is an important one and a wrongful rejection of H_0 will have serious consequences we could adopt a more stringent level, say $\alpha = 0.01$ or even $\alpha = 0.001$.

Decision Errors

When we come to stage 6 of the process above, there are two types of error we can make. The probability of making each of these types of error depends on the decision rule we adopt in 4. These types of incorrect decision are usually called type I and type II errors.

- A Type I error is the rejection of the hypothesis H_0 when it is in fact true.
- A Type II error is the acceptance of the hypothesis H_0 when it is in fact false.

These decisions are illustrated in table 10.1.

Table 10.1

	True state of the world	
Decision on H_o	*H_o is true*	*H_o is false*
Accept H_o	Correct decision	Incorrect decision i.e. Type II error
Reject H_o	Incorrect decision i.e. Type I error	Correct decision

To illustrate these types of error we might consider the hypothesis that a person is sick, i.e. H_0 = person is ill H_1 = person is not ill. The person may, in fact, be sick or not. When a doctor confronts this person in the surgery he or she has to decide whether the person is sick or not. Different decisions by doctors may have very different consequences for the person. The doctor may make a type I error and judge the person healthy when he or she is in fact ill, or a type II error of judging the person ill when he or she are in fact healthy. Doctors are trained to be cautious in diagnosis and generally see a type I error as being the most serious type. Even patients with a long record of hypochondria will be ill sometime. There may be no serious consequences of a type II error, especially if a second test is used before radical treatment such as surgery or labelling as a schizophrenic and committal to a mental hospital. So doctors are understandably reluctant to judge people healthy who present themselves at the surgery claiming to be sick. That is, doctors initially use the null hypothesis – this person is sick – and will adopt a rigorous level of significance α for a decision to reject that hypothesis. What we have called the level of significance α is in fact the chance of making a type I error, i.e. the probability of rejecting H_0 when H_0 is true.

Let us look at a further example of hypothesis testing where a two-tail test is appropriate. We again use a significance level α of 0.05.

Example 6 A scale for measuring adult authoritarian attitudes has been found over a long period of time to have a mean of 100. A random sample of 36 Merchant Navy mates and engineers was found to have a mean score of 103 on this scale, with a standard deviation of 12. Can it be concluded on the basis of this sample that they have a different level of authoritarian attitude to that found in the adult population at large?

We follow the same steps as previously.

1 State the hypothesis to be tested and its alternative

H_0: $\mu = 100$

H_1: $\mu \neq 100$

The appropriate hypothesis is the null hypothesis H_0, that μ is 100 and, as we are asked about difference, the alternative hypothesis H_1 must be in the form shown. This gives rise to a two-tail test (figure 10.15) as we are interested in differences, i.e. to say more *or* less.

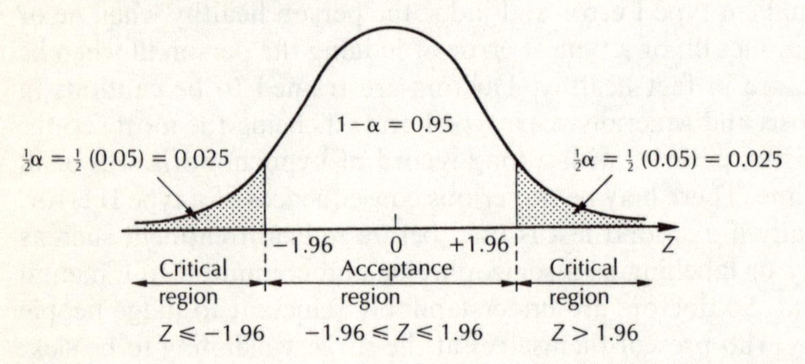

Figure 10.15

2 State the test statistic to be used and its probability distribution We will use

$$\frac{\bar{X}_1 - \mu}{S_1/\sqrt{N}} = Z$$

as a test statistic. It has a standard normal distribution.

3 Choose the value of the significance level α and determine the corresponding critical regions

let $\alpha = 0.05 \quad (P) = 0.95$.

As we now have a two-tail test our critical region of 5% is split between the tails. Looking up 0.025 in the body of Table 1, Appendix 2 we find that this corresponds to a Z value of -1.96. That is, the limiting values of the critical region are $Z = \pm 1.96$. You may remember that earlier, in chapter 9, we took this Z value to be approximately 2.

This stage of the process of hypothesis testing with two tails should also recall to mind our work in the previous chapter on confidence intervals.

4 State the decision rule for the acceptance or rejection of H_0

reject H_0 if $Z < -1.96$ or $Z > 1.96 \qquad \alpha = 0.05\ (P = 0.95)$
accept H_0 if $-1.96 \leqslant Z \leqslant 1.96$

5 Calculate the value of the test statistic on the assumption that H_0 is true

$$\text{test statistic } Z = \frac{\bar{X}_1 - \mu}{S_1/\sqrt{N}} = \frac{103 - 100}{12/\sqrt{(36)}} = 1.5$$

6 Make the decision to accept or reject H_0 In this case the value of the test statistic lies in the acceptance zone. We cannot reject H_0 on the basis of this evidence. It still stands. According to the evidence from this sample and at this level of significance ($\alpha = 0.05$), the measured authoritarianism of this group does not seem to be significantly different from that of the parent population.

Before leaving hypothesis testing we look briefly at single samples where the number in the sample is low, i.e. where there are less than 30 cases in the random sample.

Student's t Test for Small Samples

In the early part of this century it was found that the use of the Z score with samples of less than 30 measurements often led to incorrect decisions. When the sample size is less than about 30 it is not safe to assume that the shape of the sampling distribution of means is near normal. Ideally samples should be 30 or more in size but in the real world we frequently, for one reason or another, have to deal with samples which are smaller than this. A statistician working for Guinness Breweries called W. S. Gossett but writing under the pen name of Student developed a test around what he called the t distribution. This is a very useful test which is similar in application to our previous Z test but can be used for small random samples of less than 30 measurements. Unlike the Z test, however, we have to use a quantity called the degrees of freedom of the sample in order to look up the table of critical values. (Table 3, Appendix 2). The degrees of freedom are readily found by subtracting one from the number in the sample N, i.e. the degrees of freedom are equal to $N - 1$.

For a sample from a population with a normal distribution the quantity that we have called Z approximately follows the Student t distribution with $N - 1$ degrees of freedom. When N is 30 or more the Z and t distributions are virtually identical for most practical purposes. However, for small samples of less than 30 there is divergence and it is appropriate to use t rather than Z as a test statistic. However unlike the Z test we need to make the assumption that the parent population from which we draw the sample is approximately normal in shape. If we cannot make this assumption of normality then the t-test cannot be used and we would have to turn to less powerful tests such as the Mann-Whitney test (not covered in this introductory book) which do not rely on the assumption of population normality.

The t test statistic can be calculated from the following formula in much the same way that we used for Z but here we need to utilize the degrees of freedom, df.

$$\text{test statistic } t = \frac{\bar{X} - \mu}{S_1/\sqrt{N}}$$

We calculate our test statistic value from the above formula and compare it with the critical values of the t distribution to be found in Neave, Table 3 Appendix 2. This table is entered with the chosen significance level, e.g. $\alpha = 0.05$, and the number of degrees of freedom df = $N - 1$ to get the required critical value for t. As with the Z distribution we decide whether the observed test statistic lies inside or outside the rejection zones bounded by the values found in the table.

Let us now look at an example using the t statistic. We utilize the same six steps as previously.

Example 7 A new method of teaching touch typing was employed with a class of 10 typing students. At the end of the test period they had acquired the following speeds in words per minute:

15.0 16.6 18.2 17.4 21.8 12.2 12.6 19.0 23.0 21.8

It is known from experience with a large number of typists using an old standard method of teaching that an average speed of 20 words per minute was achieved in the same period and that the speeds where normally distributed. Do the data above indicate that the new method makes any difference to learning speeds?

As we have a small sample of 10 from a normal population it is appropriate to use a t test and as we are interested in a difference either way we will use a two-tail test. We proceed as before.

1 State the hypothesis to be tested and its alternative

H_0: $\mu = 20$
H_1: $\mu \neq 20$

2 State the test statistic to be used and its probability distribution

$$t = \frac{\bar{X}_1 - \mu}{S_1/\sqrt{N}}$$

The test statistic has a t distribution.

3 Choose the value of the significance level α, and determine the critical region. Significance level $\alpha = 0.05$ ($P = 0.95$). Two-tail test with degrees of freedom df $= N - 1 = 9$ (figure 10.16).

Critical values from Table 3 Appendix 2. $t = \pm 2.2622$

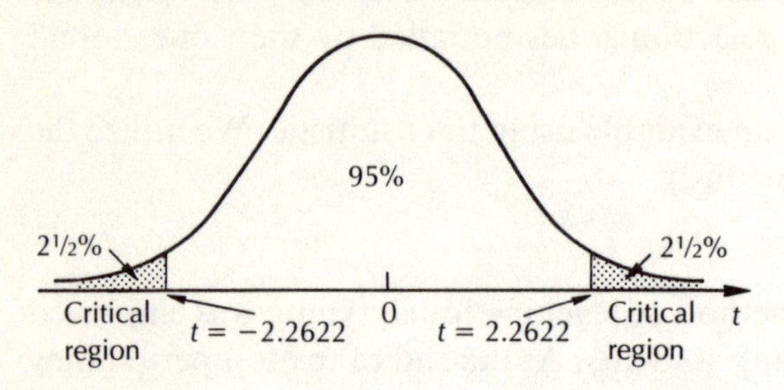

Figure 10.16

4 State the decision rule for the acceptance or rejection of H_0

reject H_0 if our observed t value is less than or equal to -2.2622

reject H_0 if our observed t value is greater than or equal 2.2622

5 Calculate the value of the test statistic t on the assumption that H_0 is true

$$t = \frac{\bar{X}_1 - \mu}{S_1/\sqrt{N}} = \frac{(17.76 - 20)}{3.778/\sqrt{10}} = -1.875$$

Sample mean $\bar{X}_1 = 17.76$; sample standard deviation $S_1 = 3.778$; sample number $N = 10$; degrees of freedom $N - 1 = 9$.

6 Make the decision to accept or reject H_0 Our observed t value of -1.875 is fairly low but it does not lie in either of the critical regions and so we cannot on the basis of this sample reject the null hypothesis H_0. It still stands. From this sample it seems that the new method of teaching touch typing neither improves or impedes typing speeds significantly.

Postscript

In this chapter we have seen how we can transform normal or near-normal distributions to the standard normal form. This allowed us to use normal tables such as those given by Henry Neave to relate Z scores, areas and probabilities. We also looked at hypothesis testing in relation to population means using two test statistics Z and Student's t.

In the next chapter we look at a distribution derived from the normal distribution, i.e. the χ^2.

Exercises

1 Find the following probabilities for the Standard Normal variable Z:

(i) $P(-1 \leqslant Z \leqslant 1.5)$
(ii) $P(-0.3 \leqslant Z \leqslant 1.2)$
(iii) $P(0.4 \leqslant Z \leqslant 2.4)$
(iv) $P(-2 \leqslant Z \leqslant -1)$
(v) $P(2 \leqslant Z \leqslant 3)$
(vi) $P(Z \leqslant 1.645)$
(vii) $P(Z > 1.645)$
(viii) $P(Z > -1.5)$
(ix) $P(Z \leqslant -1.5)$
(x) $P(-0.674 \leqslant Z \leqslant 0.674)$

2 In a test of reading attainment for 10-year-old children it was found that the mean score for a large number of children was 100 and the standard deviation 10. If the scores were distributed in a near-normal fashion, what proportion of the children taking this test are likely:

(i) to score more than 100?
(ii) to score more than 90?
(iii) to score between 95 and 105?
(iv) What is the approximate probability that a child selected at random from this population will have a reading attainment score over 120?

3 The birth-weights X of a large number of female babies at a certain Welsh hospital over a one-year period were found to be near-normal in distribution with a mean weight of 3100 g and a standard deviation of 380 g.

(i) Evaluate the probability $P(2500 \leqslant X \leqslant 3500)$.
(ii) Find the proportion of underweight babies, i.e. $P(X \leqslant 2500)$
(iii) Within what weight limits would you expect the middle 95% of these baby weights to lie?

4 The manufacturer of Brand X light bulbs claims that their mean life is 1600 hours. However, a random sample of 100 such bulbs was found by a consumer organization to have a mean life of only 1580 hours with a standard deviation of 60 hours. Test the hypothesis that the mean life of Brand X bulbs is less than 1600 hours using a significance level of 0.05.

5 In a large number of maze-running experiments in a psychological laboratory it was found that mature male rats completed sets of runs in the maze with a mean time of 22 seconds when rewarded with food pellets on reaching the end. For a randomly selected group of 30 mature male rats the maze was run under the same conditions except that the reward (stimulus) was not food but being able to mount (momentarily) a receptive female rat. Their mean time to get through the maze was 19 seconds with a standard deviation of 8 seconds. Can it be concluded that the change of stimulus makes the learning time significantly different? (Test at $\alpha = 0.05$.)

6 A new coaching method in primary mathematics was tried with a group of 10-year-old children. In a standard test they achieved the following percentage marks:

40 43 46 47 55 59 63 63 66 72 73 75 80 82 ($N = 14$)

It is known that the scores for this type of test are normally distributed and the mean score for a large number of pupils is 55%. Use a t test to determine whether the above group of children did significantly better in the test than the mean score of 55%. (Test at $\alpha = 0.05$.)

Answers to Exercises

1 (i) $P(-1 \leqslant Z \leqslant 1.5) = \Phi(1.5) - \Phi(-1) = 0.9332 - 0.1587 = 0.7745$

(ii) $P(-0.3 \leqslant Z \leqslant 1.2) = \Phi(1.2) - \Phi(-0.3) = 0.8849 - 0.3821 = 0.5028$

(iii) $P(0.4 \leqslant Z \leqslant 2.4) = \Phi(2.4) - \Phi(0.4) = 0.99180 - 0.6554 = 0.3364$

(iv) $P(-2 \leqslant Z \leqslant -1) = \Phi(-1) - \Phi(-2) = 0.1587 - 0.0228 = 0.1359$

(v) $P(2 \leqslant Z \leqslant 3) = \Phi(3) - \Phi(2) = 0.00865 - 0.9772 = 0.02145$

(vi) $P(Z \leqslant 1.645) = \Phi(1.645) = 0.95$

(vii) $P(Z > 1.645) = 1 - \Phi(1.645) = 1 - 0.95 = 0.05$

(viii) $P(Z > -1.5) = 1 - \Phi(-1.5) = 1 - 0.0668 = 0.9332$

(ix) $P(Z \leqslant -1.5) = \Phi(-1.5) = 0.0668$

(x) $P(-0.674 \leqslant Z \leqslant 0.674) = \Phi(0.674) - \Phi(-0.674) = 0.750 - 0.250 = 0.5$

2 $X =$ reading score $\quad \mathrm{Z} = \frac{\mathrm{X}-\mu}{\sigma} \quad \mu = 100 \quad \sigma = 10$

(i) $P(X > 100) = P\left(\frac{X-100}{10} > \frac{90-100}{10}\right)$

$= P(Z > 0) = 0.5$

(ii) $P(X > 90) = P\left(\frac{X-100}{10} > \frac{90-100}{10}\right) = P(Z > -1)$

$= 1 - P(Z \leqslant -1) = 1 - \Phi(-1) = 1 - 0.1587 = 0.8413$

(iii) $P(95 \leqslant X \leqslant 105)$

$= P\left(\frac{95-100}{10} \leqslant \frac{X-100}{10} \leqslant \frac{105-100}{10}\right)$

$= P(-0.5 \leqslant Z \leqslant 0.5) = \Phi(0.5) - \Phi(-0.5)$

$= 0.6915 - 0.3085 = 0.383$

(iv) $P(X > 120) = P\left(\frac{X-100}{10} > \frac{120-100}{10}\right) = P(Z > 2)$

$= 1 - P(Z \leqslant 2) = 1 - \Phi(2)$
$= 1 - 0.9772 = 0.0228$

3 X = birthweight $\mu = 3100$ g $\sigma = 380$ g

(i) $P(2500 \leqslant X \leqslant 3500)$

$$= P\left(\frac{2500 - 3100}{380} \leqslant \frac{X - \mu}{\sigma} \leqslant \frac{3500 - 3100}{380}\right)$$

$$= P(-1.58 \leqslant Z \leqslant 1.05) = \Phi(1.05) - \Phi(-1.58)$$

$$= 0.8531 - 0.0571 = 0.796$$

That is, about 80% of these babies are within the weight limits of 2500–3500 g.

(ii) $P(X \leqslant 2500) = P\left(\frac{X - \mu}{\sigma} \leqslant \frac{2500 - 3100}{380}\right)$

$$= P(Z \leqslant -1.58) = 0.0571$$

That is about 6% of these babies were of low birthweight and so potentially at risk.

(iii) The middle 95% of these babies would lie in the area bounded by $Z = \pm 1.96$ on the standard normal curve, i.e.

$$P(-1.96 \leqslant Z \leqslant 1.96) = 0.95$$

We can get the X values from this by transposing the formula

$$Z = \frac{X - \mu}{\sigma}$$

to get $X = \mu + Z\sigma$. Substitue the two Z values of -1.96 and 1.96 in this equation to get the corresponding X values

$$X_1 = -1.96 \times 380 + 3100 = 2355g$$

$$X_2 = 1.96 \times 380 + 3100 = 3845g$$

If you do not like this method you can argue that, as the middle 95% of the area is within 1.96 standard deviations of the mean of the Standard Normal curve (standard deviation, 1), the same proportion of the area must lie within 1.96 standard deviations of the normal distribution of weights, i.e.

$$\pm 1.96\sigma = 3100 \pm 1.96 \times 380$$
$$= 3100 \pm 745 = 2355 \text{ to } 3845 \text{ g}$$

4

1 Hypothesis to be tested and its alternative:

H_0: $\mu = 1600$
H_1: $\mu < 1600$

2 Test statistic:

$$\frac{\bar{X}_1 - \mu}{S_1/\sqrt{N}} = Z$$

3 Significance level: $\alpha = 0.05$, using a one-tail test, $Z = -1.645$.

4 Decision Rule:

reject H_0 if $Z \leqslant -1.645$
accept H_0 if $Z > -1.645$

5 Value of test statistic:

$$Z = \frac{\bar{X}_1 - \mu}{S_1/\sqrt{N}} = \frac{1580 - 1600}{6/\sqrt{(100)}} = -3.33$$

6 Decision: Our observed test statistic $Z = -3.33$ is less than -1.645 and so falls in the critical zone. On the basis of this sample we must reject H_0. At this level it does appear that the life of Brand X light bulbs is significantly less than 1600 hours.

5

1 Hypothesis to be tested and its alternative:

H_0: $\mu = 22$
H_1: $\mu \neq 22$

2 Test statistic:

$$Z = \frac{\bar{X}_1 - \mu}{S_1/\sqrt{N}}$$

3 Significance level: $\alpha = 0.05$ and a two-tail test, $Z = \pm 1.96$. We should note here that we are asked whether the learning time is significantly different. No direction is implied here (e.g. faster) so we should use a two-tail test. Had we been asked if the change of stimulus made the rats faster, we would have used a one-tail test.

4 Decision rule:

reject H_0 if $Z < -1.96$ or $Z > 1.96$
accept H_0 if $-1.96 \leqslant Z \leqslant 1.96$

5 Value of test statistic:

$$Z = \frac{\bar{X}_1 - \mu}{S_1/\sqrt{N}} = \frac{19 - 22}{8/\sqrt{39}} = -2.05$$

6 Decision: Our test statistic $Z = -2.05$ is less than -1.96 and therefore lies in the critical zone. So we must reject H_o. It does appear that the new stimulus is different from the old in its effect on the mean time of runs. The learning time for male rats under these new conditions is significantly different and appears to be less.

6

1 Hypothesis to be tested and its alternative:

H_0: $\mu = 58$
H_1: $\mu > 58$

2 Test statistic:

$$t = \frac{\bar{X}_1 - \mu}{S_1/\sqrt{N}}$$

3 Significance level: $\alpha = 0.05$, using a one-tail test; degrees of freedom $= N - 1 = 13$; critical value from Neave $t = 1.77$.

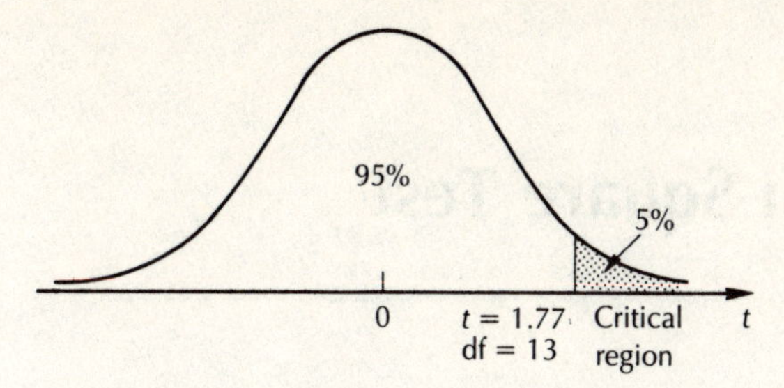

4 Decision rule: Reject H_0 if our observed value of $t \geqslant 1.77$.

5 Value of test statistic:

$$t = \frac{\bar{X}_1 - \mu}{S_1/\sqrt{N}} = \frac{61.7 - 55}{13.9/\sqrt{14}} = 1.803$$

Sample mean $\bar{X}_1$ 61.7; sample standard deviation S_1 13.9; number N in sample, 14.

6 Decision: Our observed value of t is 1.803 and it does lie in the critical region. On the basis of this sample we must reject the null hypothesis H_0. The new coaching method does seem to improve the test score significantly.

11 The Chi-Square Test

In this final chapter we turn from the normal distribution which has been the focus of our attention in the last two chapters to its close relative, the χ^2 distribution (pronounced ki-square with the k as in kite). We examine how this probability distribution can help us make decisions about tabulated categorical data, i.e. about data that fall into two or more categories and can be expressed in the form of a table.

We look briefly at the χ^2 as a type of probability distribution but we will concentrate more on its use in helping us decide whether any pattern to be found in the cells of the table relating to the two variables is due to chance or not.

Let us first look at the type of problem we shall be addressing; we then say a bit about χ^2 as a probability distribution before looking at its utility in helping us make decisions about hypotheses.

Example 1 The mean number of books borrowed per week from a specific but obscure section of a university library over a term was as follows.

	Mon	*Tues*	*Wed*	*Thurs*	*Fri*
mean no. of books borrowed	13	9	12	11	15

Use the χ^2 test to examine the hypothesis that the number of books borrowed depends on the day of the week.

In this example there are two categorical variables, day of the week and number of books borrowed. The question asks whether these two variables are independent. We wish to find out more about the relationship between these variables or whether the pattern we see in the distribution is due to chance.

Let us for a moment think back to the procedure we adopted in the last chapter for testing hypotheses about means using Z tests. On the basis of a null hypothesis we decided on a test statistic, Z, and then calculated the Z score and tested whether such a score was significant by relating it to the distribution of all such Z scores, i.e. to their theoretical sampling distribution. Using the central limit theorem, we assumed that the theoretical distribution of all such randomly chosen Z scores approached a normal distribution when we took a large number of them. We then, with our knowledge of the characteristics of the normal curve, estimated where in this sampling distribution of means our sample Z score was likely to fall.

As we saw, the fundamental basis of such statistical inference is the sampling distribution of the test statistic against which we compare our randomly selected sample test statistic. We will proceed in a similar fashion here except that our single test statistic will now be χ^2 rather than Z and the appropriate sampling distribution of the test statistic follows a χ^2 rather than a normal distribution. Each χ^2 test statistic has a characteristic sampling distribution which depends on the sample size and a quantity called the degree of freedom which we met in the last chapter in relation to the t distribution for small samples.

In that it has the added dimension of degrees of freedom the sampling distribution of the test statistic χ^2 is more complex than the standard normal distribution we used in the case of the Z test statistic. However, as we shall see, the sampling distribution of χ^2 is closely linked to the standard normal distribution.

Here we need to note that we have our test statistic which I will call calc. χ^2 (short for calculated χ^2) and as before we will compare this with a critical value of the sampling distribution of χ^2 for the correct degree of freedom. Characteristics of such sampling distributions of χ^2 for various degrees of freedom are to be found in appendix 2 Table 4. I will call these theoretical sampling distribution values tab. χ^2 (short for tabulated χ^2). They are the critical values against which we will measure the test statistic.

As with the Z score we can use different levels of probability to compare calc. χ^2 with tab χ^2; however, for simplicity in this section we will generally use the 0.05 significance level ($\alpha = 0.05$, $P = 0.95$).

However, before looking closer at the test statistic, calc. χ^2 and the sampling distribution of χ^2, tab. χ^2 (critical values), we need to look at the Null hypothesis and the concept of degrees of freedom.

The Null Hypothesis H_0

As in the previous chapter on the testing of hypotheses we formulate the hypothesis in the null or 'no difference' way. That is, we will always state that the two variables in the table (e.g. number of books borrowed and day of the week) are independent. In the case of our example this would amount to saying that the day of the week makes no difference to the number of books taken out. Such differences that are observed, under this null hypothesis, are due to random chance and not to an underlying systematic pattern. As noted in the last chapter we use this null form of hypothesis even if we are fairly sure, and indeed it is our wish to demonstrate, that there is a difference between the probabilities of the various outcomes.

As we shall see, stating the hypothesis in this negative way rather than a positive way makes it easier to calculate the expected values (expected values if the H_0 is true) for the cells of the distribution. We could also note in passing that it is easier to prove things false or untrue than to show them to be correct or true. Truth, in this sense, is not symmetrical. The alternative hypothesis H_1 will state that the two variables are not independent. This is equivalent to saying there is some relationship between them.

Degrees of Freedom

We came across the term degrees of freedom in the last chapter in relation to Student's t distribution, where we found its value by subtracting one from the number in the sample. Here we do something similar. The concept is rather complex but we can get some appreciation of it by considering the number of boxes or cells into which the distribution of probabilities can fall. The term degrees of freedom refers to the number of ways in which the boxes or cells can be independently filled. It is important because it determines the shape of the various sampling distributions of χ^2. We should note that with the Z test for means we only had one sampling distribution, the standard normal distribution, in which we wished to 'place' our test statistic; here, however, we are dealing with a whole family of probability sampling distributions. Each sampling distribution varies in shape according to the number of degrees of freedom as shown later.

The concept can be illustrated as follows. Suppose there are three possible ways in which an event could occur, i.e. there are three boxes or cells

cell 1	cell 2	cell 3

into which the observations could fall and we are dealing with say 10 occurrences of the event. Even if we know how many of these 10 events fall in the first cell we are still unsure how cells 2 and 3 will be filled up. If, on the other hand, we do know how many of the 10 events have fallen in cell 1 and 2 there will be no uncertainty about how many there will be in cell 3. Knowing the total number of events we are dealing with and the number of these already in cells 1 and 2 we can know, by simple subtraction, what must go in cell 3. For instance if we know that four of the events have fallen in cell 1, and five in cell 2, it must follow that cell 3 contains only one event ie $10 - (4 + 5) = 1$. The total number of events is 10.

4	5	?

That is, in this case there are only 2 degrees of freedom (df = 2) because, once we know the way any two cells are filled, then the contents of the third cell are completely determined.

So if we have only one row (or indeed column) of cells we can find the number of degrees of freedom easily, as we did with the *t* distribution, by subtracting one from the number of cells in the row (or column), i.e. df = $n - 1$. For our library example we have df = $n - 1 = 5 - 1 = 4$. For tables with more than one row (or column) there is a simple rule we can use to find the degrees of freedom.

If we have a table or matrix of numbers with *n* cells in the rows and *m* cells in the columns the degrees of freedom are given by df = (n − 1) (m − 1). For example, if we have a 3 × 4 matrix or table, say 3 rows and 4 columns as shown, the degrees of freedom will be given by

e_1	e_2	e_3		T_5
e_4	e_5	e_6		T_6
T_1	T_2	T_3	T_4	

$$
\begin{aligned}
\text{df} &= (n - 1)(m - 1)\\
&= (3 - 1)(4 - 1)\\
2 \times 3 &= 6
\end{aligned}
$$

That is, once six cells have been filled as indicated ($e_1, e_2 \ldots e_6$) and we have the row and column totals ($T_1, T_2 \ldots T_6$) the numbers in the remaining cells will be completely determined.

We now look at the test statistic and the sampling distribution.

The Test Statistic – calc. χ^2

The test statistic which I call calc. χ^2 is calculated by comparing the observed cell frequencies with the frequencies we could expect in the cells under the null hypothesis H_0. Let us illustrate this by using our library example. The appropriate null hypothesis here is that the two variables (day of the week and number of books borrowed) are independent, i.e. the day of the week makes no difference to the number of books borrowed. We are dealing with a total of 60 books borrowed and so under the 'no difference' hypothesis H_0 we would expect them to be equally distributed across the 5 days with 12 books in each cell. These are our expected (e) cell values under the null hypothesis and we are now in a position to compare them with the observed (o) cell values.

	Mon	*Tues*	*Wed*	*Thurs*	*Fri*
observed (o)	13	9	12	11	15
expected (e)	12	12	12	12	12

Our test statistic calc. χ^2 compares corresponding cells and sums the squared differences. The squaring of differences is done to enable us to sum both positive and negative differences. The squared differences for each pair of cells is divided by the expected cell number so that, if necessary, we can compensate for large cell numbers which might give rise to disproportionately large differences.

The test statistic calc. χ^2 is defined as

$$\text{calc.}\chi^2 = \sum \frac{(o-e)^2}{e}$$

We can compute calc. χ^2 for our distribution as follows:

$$\begin{aligned}\text{calc.}\chi^2 &= \frac{(13-12)^2}{12} + \frac{(9-12)^2}{12} + \frac{(12-12)^2}{12} + \frac{(11-12)^2}{12} \\ &\quad + \frac{(15-12)^2}{12} \\ &= 1.67\end{aligned}$$

This is the value of our test statistic and now, as with our Z score procedure, we need to try and 'place' this value in the appropriate sampling distribution to see whether it is significant. That is to say we need to compare our test statistic calc. χ^2 with the appropriate critical value tab. χ^2. Let us now look at the sampling distribution of χ^2 , i.e. tab. χ^2.

The Sampling Distributions of χ^2 – tab. χ^2

This is an introductory text and it is only appropriate here to try and illustrate this concept and get some feel for its use.

The family of χ^2 distributions each with its own degree of freedom is related to the standard normal distribution we spent much time with in the last two chapters. It is defined as:

$$\text{tab. } \chi^2 = \left(\frac{X_1 - \mu}{\sigma}\right)^2 + \left(\frac{X_2 - \mu}{\sigma}\right)^2 + \left(\frac{X_3 - \mu}{\sigma}\right)^2 + \ldots + \left(\frac{X_n - \mu}{\sigma}\right)^2 = Z_1^2 + Z_1^2 + Z_3^2 + \ldots + Z_n^2 = \sum_1^n Z^2$$

That is, tab. χ^2 is the Z^2 values summed from 1 to n. X_1, X_2, X_3, ..., X_n is a random sample drawn from a normal population with mean μ and standard deviation σ. The Zs are, as before, the Z scores. That is, the tab. χ^2 distribution with $n - 1$ degrees of freedom is the sum of the squares of n independent standard normal random variables Z_n.

As mentioned previously, each of these χ^2 sampling distributions has a distinctive shape based on its degrees of freedom. Figure 11.1 gives the approximate shape of a few sampling distributions of χ^2 under the assumption that the null hypothesis is true. There are at least three points to note about such probability distributions

1 Unlike the distribution of Z scores that we looked at in chapter 10, χ^2 values are always positive (the squaring of the differences ensures this).
2 Each of the sampling distributions has a characteristic shape and these curves, especially the 'early' ones in the family (those with low degrees of freedom.) are distinctly skewed to the right (lopsided in shape with long tails to the right). That is, for relatively low degrees of freedom, up to say df = 8 we have distinct skews to the right, but after this as the degrees of freedom increase the distributions become more and more bell shaped.

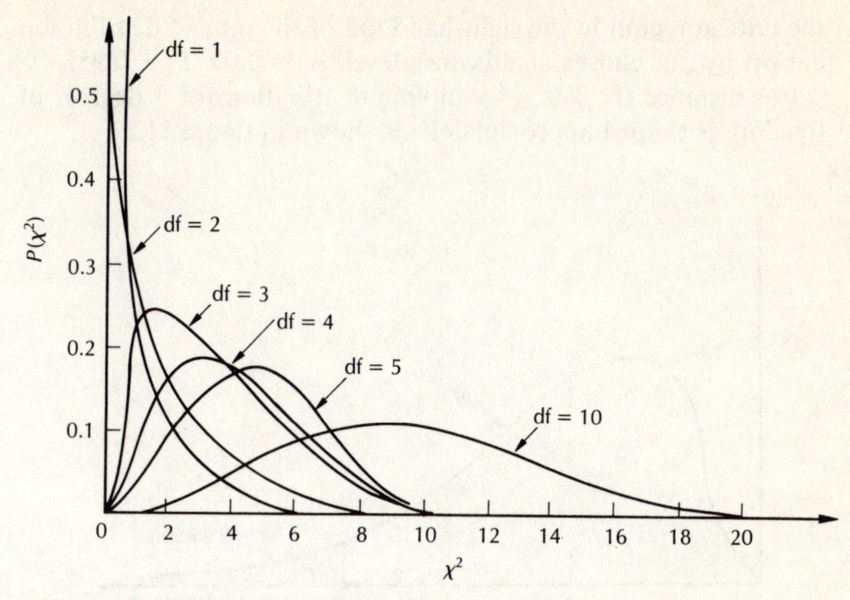

Figure 11.1 The distribution of χ^2 for various degrees of freedom

3 It can be shown that the expected value (the mean value, i.e the value with the greatest likelihood of being observed) of χ^2 is close to its number of degrees of freedom. Thus for a χ^2 distribution with say V degrees of freedom the mean will be approximately V. Its standard deviation is also related to the number of degrees of freedom and is equal to $\sqrt{(2V)}$.

As the degrees of freedom exceed about 10 the sampling distribution becomes more symmetrical and bell shaped and its median, mode and mean come closer together.

In relation to point (3) we could expect a test statistic of χ^2 of about 4, i.e. calc. $\chi^2 \approx 4$, to occur fairly frequently in our library example as the degrees of freedom are, as we have seen, 4. A value of say 8 would be extremely rare in our library example and probably highly significant if it arose when the degrees of freedom were only 4. However, if we had a test statistic of 8 arising from a 3×5 table (df $= (3 - 1) \times (5 - 1) = 2 \times 4 = 8$) this could be relatively common and probably not significant.

The test statistic calc. χ^2 is measured against the critical value tab. χ^2 with the appropriate degrees of freedom and the values of calc. χ^2 that we will consider to be significant are those which lie in the critical region in the right-hand tail of the tab. χ^2 distribution cut off by our chosen significance level ($\alpha = 0.05$, $P = 0.95$).

For instance the tab. χ^2 sampling distribution for 4 degrees of freedom is shaped approximately as shown in figure 11.2.

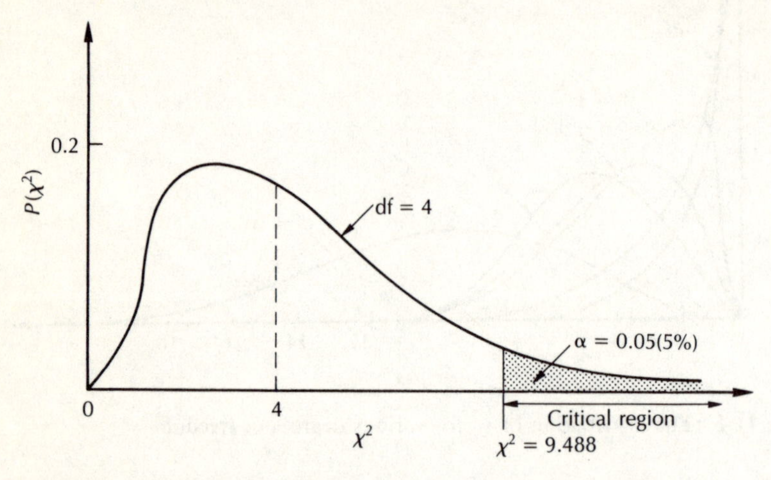

Figure 11.2 The distribution of χ^2 for four degrees of freedom

There is only a probability of about 0.05 (5%) that the test statistic calc. χ^2 could have a value greater than 9.488 by random chance (see Table 4, Appendix 2). If our test statistic calc. χ^2 is big enough to lie in the critical region (calc. $\chi^2 \geqslant 9.488$) we will reject the null hypothesis that the two variables being examined (day of the week and number of books borrowed) are independent. We will judge values of calc. χ^2 of less than this (calc. $\chi^2 < 9.488$) not to be significant. From what we have said previously we would expect values of the test statistic of about 4 to occur relatively frequently as df = 4.

In general terms the larger the value of calc. χ^2 the more we will distrust the null hypothesis H_0. This is easily appreciated by remembering that calc. χ^2 is calculated by comparing the observed and expected values and if the expected values, based on H_0, and the observed values completely coincide then calc. χ^2 will be zero because the difference between them, the quantity $o - e$, will be zero for each pair of cells. Therefore it follows that the further we move from zero for calc. χ^2 the more evidence we will have to suspect and reject the validity of the null hypothesis H_0.

At the end of this chapter we return to these sampling distributions of χ^2 and simulate a few with different degrees of freedom using the basic formula for tab. χ^2 given above (p. 328). However, with our new knowledge of the test statistic calc. χ^2 and the sampling distribution tab. χ^2 let us now return to our library example, lay out the procedure in an orderly fashion, test our hypothesis and come to a decision.

We adopt similar steps to those we used in our Z test of means.

1 State the hypothesis to be tested: H_0: the two variables, day of the week and number of books borrowed, are independent. There is no difference in the expected number of books borrowed each day and any observed differences are only what might be expected by chance. H_1: the two variables are not interdependent.

2 State the test statistic: The test statistic will be taken to be calc. χ^2 where

$$\text{calc.}\chi^2 = \sum \frac{(o-e)^2}{e}$$

where o are the observed values and e are the expected values calculated on the assumption that the null hypothesis H_0 is true.

3 Choose the value of the significance level α: Let $\alpha = 0.05$ ($P = 0.95$).

4 State the sampling distribution and critical value: The sampling distribution will be tab. χ^2 with $(n-1)=5-1=4$ degrees of freedom (df=4). Critical value 9.488 (Table 4 Appendix 2)

5 State the decision rule for the acceptance or rejection of H_0: With a significance level of $\alpha = 0.05$ and df $= 4$ we will

reject H_0 if calc. $\chi^2 \geqslant 9.488$
accept H_0 if calc. $\chi^2 < 9.488$

6 Calculate the value of the test statistic calc. χ^2 on the assumption that H_0 is true:

$$\text{calc. } \chi^2 = \sum \frac{(o - e)^2}{e} = 1.67$$

7 Make the decision to accept or reject H_0: Our test statistic calc. χ^2 = 1.67 does not lie in the critical region, i.e. greater than or equal to 9.488, and so we cannot on the basis of this information reject H_0. It still stands. On the basis of this sample it does appear that the two variables, day of the week and the number of books borrowed, are independent.

Let us now look at an example which has a 2 x 2 table.

Example 2 In a study of funerals (Foster, 1989) it was found that 887 bodies were disposed of as in table 11.1 according to social class. Can it be concluded from the figures that social class and type of funeral are independent variables?

Table 11.1 Funeral management

	Social Class	
Type of Funeral	*Non-manual*	*Manual*
Cremation	74%	64%
Burial	26%	36%
	100% (436)	100% (451)

Source: Foster 1989

The first thing to note is that we cannot apply our χ^2 technique directly to table 11.1 as the percentages in the two columns are calculated from different bases. With tables of rates, proportions or percentages we must work our way back to the basic figures, in this case bodies. In the first cell 74% represents $74 \times 436/100$ = 322.64 = 323 bodies. Similarily the other percentages give observed figures and totals as follows:

observed			
	323	289	612
	113	162	275
	436	451	887

We can now proceed as previously.

1 State the hypothesis to be tested H_0: the two variables of social class and funeral arrangements are independent. H_1: the two variables of social class and funeral arrangements are not independent.

2 State the test statistic: The test statistic will be

$$\text{calc.}\chi^2 = \sum \frac{(o-e)^2}{e}$$

3 Choose the value of the significance level α*:* Let $\alpha = 0.05$ ($P = 0.95$).

4 State the sampling distribution and critical value: The sampling distribution will be tab. χ^2 with $(n - 1)(m - 1) = (2 - 1)(2 - 1) = 1$ df
Critical value 3.841 (Table 4, Appendix 2)

5 State the decision rule for the acceptance or rejection of H_0*:* With a significant level $\alpha = 0.05$ and df $= 1$

reject H_0 if calc. $\chi^2 \geqslant 3.841$
accept H_0 if calc. $\chi^2 < 3.841$

6 Calculate the value of the test statistic calc. χ^2 *on the assumption that* H_0 *is true:* We now need to work out the expected values for the cells assuming H_0 is true. We cannot alter the row or column totals but we can fill the cells assuming that what goes in them is related not to social class or type of arrangement but only to the proportions indicated by the subtotals and total.

a	b	612
c	d	275
436	451	887

For example, what goes in cell a, if H_0 is true, will depend on the proportion of cremations (612/887) and the number of non-manual bodies to be disposed of (i.e. 436). Therefore

$$a = 436 \times \frac{612}{887} = 300.8$$

The other cell values *b*, *c* and *d* can now be found by subtraction as there is only 1 degree of freedom and we have used that in calculating cell *a*. What must go in cells *b*, *c* and *d* is completely determined by the value of *a* and the subtotals. However, it is usually wise to compute one more cell value directly as a check.

$$d = 451 \times \frac{275}{887} = 139.82$$

We now have observed and expected values as follows. I have rounded the expected values for ease of presentation.

o			*e*		
	323	289		301	311
	113	162		135	140

$$\begin{aligned}\text{calc. } \chi^2 &= \sum \frac{(o-e)^2}{e} \\ &= + \frac{(323-301)^2}{301} \quad \frac{(289-311)^2}{311} \\ &\quad + \frac{(113-135)^2}{135} + \frac{(162-140)^2}{140} \\ &= 10.21\end{aligned}$$

7 Make the decision to accept or reject H_0: The test statistic calc. χ^2 is 10.21. This is greater than tab. χ^2 of 3.841 (α = 0.05, df = 1) and therefore by our decision rule (5) we must reject the null hypothesis H_0. The two variables of social class and type of funeral do not appear to be independent in this study. Social class and funeral arrangements do appear to be related. Non manual classes appear to favour cremation.

Let us now tackle a χ^2 analysis with Minitab to help us work out the expected values and the value of calc. χ^2. In this example we also meet the common problem of small cell numbers.

Example 3 The example is to do with the struggle of women to become priests in the Church of England and comes from a Church Assembly vote reported in the *Daily Telegraph*.

The *Daily Telegraph* of 4 July 1967 carried a report (by Our Ecclesiastical Correspondent) about the Church Assembly's rejection of a motion by Miss Valerie Pitt that individual women who felt called to exercise 'the office and work of a priest in the Church' should now be considered, on the same basis as individual men, as candidates for Holy Orders.

Table 11.2 Voting in the church

	Vote		
House	*Aye*	*No*	*Abstained*
House of Bishops	1	8	8
House of Clergy	14	96	20
House of Laity	45	207	52

Source: Open University, M345 Statistical Methods, Unit 9, 1987, p. 28

The vote on the motion is shown in table 11.2. Are the two variables, voting and status of the voter (House), independent variables?

1 State the hypothesis is to be tested H_0: the two variables of voting and status of the voter (House of the Church) are independent. H_1: the two variables of voting and status of the voter are not independent.

2 State the test statistic: The test statistic will be

$$\text{calc.}\chi^2 = \sum \frac{(o - e)^2}{e}$$

3 Choose the value of the significance level α*:* Let $\alpha = 0.05$ ($P = 0.95$)

4 State the sampling distribution and critical value: The sampling distribution will be tab. χ^2 with $(n - 1)(m - 1) = (3 - 1)(3 - 1) = 2 \times 2 = 4$ df. Critical value 9.488 (Table 4 Appendix 2)

5 State the decision rule for the acceptance or rejection of H_0: With a significance level $\alpha = 0.05$ and df $= 4$

reject H_0 if calc. $\chi^2 \geqslant 9.488$
accept H_0 if calc. $\chi^2 < 9.488$

6 Calculate the value of the test statistic calc. χ^2 on the assumption that H_0 is true:

We can read the data in and proceed as follows:

```
MTB > # Chi-square analysis of data re. voting
MTB > #for women priests.
MTB > READ these data into C1-C3
DATA> 1   8   8
DATA> 14 96  20
DATA> 45 207 52
DATA> END
      3 ROWS READ
MTB > CHISQUARE test on data in C1-C3

Expected counts are printed below observed counts

            C1       C2       C3    Total
1            1        8        8       17
          2.26    11.72     3.02

2           14       96       20      130
         17.29    89.65    23.06

3           45      207       52      304
         40.44   209.63    53.92

Total       60      311       80      451

ChiSq =  0.704 + 1.182 + 8.239 +
         0.628 + 0.450 + 0.406 +
         0.513 + 0.033 + 0.069 = 12.224
df = 4
2 cells with expected counts less than 5.0
```

7 Make the decision to accept or reject H_0: The test statistic calc. $\chi^2 = 12.224$. This is greater than our critical value tab. χ^2 of 9.488 ($\alpha = 0.05$, df = 4) and therefore by our decision rule (5) we must reject the null hypothesis H_0. Voting behaviour does appear to be related to status in the Church. There are significant differences in the way the various Houses of the Church voted in this matter.

The above decision is correctly made in accordance with our procedure and decision rule (5) but before accepting it as valid we need to consider a common limitation of the χ^2 test, low cell numbers. When expected cell numbers are low most statisticians would advise caution. There is some dispute over the effect of low cell numbers and what is to count as a low cell number, but in an introductory text such as this we only need to be aware of the problem and its general solution. The solution is usually seen to lie in merging some of the categories; however, in the present case this is not a suitable solution.

Look back for a moment at the Minitab print-out and you can see that it helpfully warns us that two expected cell counts are less than 5. Where expected cell numbers fall below this arbitrary figure of 5 we are faced with the possibility that our decision rule may lead us to make an incorrect decision. In some circumstances where this happens we can merge categories to increase cell numbers; however, in the case we are considering we cannot realistically merge the category of Bishop with that of Clergy without losing much of the value of the data. The Bishops in fact constitute the most interesting category in our example. Rather than merging categories it would be better in our case to take these two low cell numbers into account and not reject our null hypothesis.

When we examine the print-out we see that one of these cells with low expected numbers contributes disproportionately to the calculated χ^2 value. (The abstaining Bishops are the greatest source of heterogeneity in the table.) This is an indication that our calculated χ^2 value which we used to reject the null hypothesis may be too high. Despite our decision rule we cannot in these circumstances prudently reject the null hypothesis – it still stands.

To turn from our particular example to the general case, we should note that there are many instances where small expected cell numbers can be usefully merged without affecting the data too much, and as a general rule categories should be merged if an expected cell value is less than 5.

Simulated Sampling Distributions of χ^2

Finally in this chapter let us look, as promised, at various sampling distributions of χ^2 (tab. χ^2) with the aid of Minitab simulations.

Earlier we saw that the sampling distribution of χ^2 (tab. χ^2) for various degrees of freedom is

$$\text{tab. } \chi^2 = Z_1^2 + Z_2^2 + Z_3^2 + \ldots + Z_n^2$$

for various values of n. Each sampling distribution is made up of squared Z scores. As can be seen from the diagram (page 329) the shape of the sampling distribution varies according to the number of degrees of freedom. For low degrees of freedom, say up to 8 or so, the change in the shape of the distribution is considerable.

We start with random observations from a standard normal distribution (mean $\mu = 0$, standard deviation $\sigma = 1$), square them and plot them in histogram form. These will form our simulated sampling distributions. We will then compare them with more directly simulated χ^2 distributions.

```
MTB > #Taking 300 random samples from a standard
MTB > #normal distribution and squaring them to
MTB > #obtain a chi-square distribution with one
MTB > #degree of freedom.
MTB > #Then taking 300 random samples from a
MTB > #chi-square distribution, df=1, and
MTB > #comparing the two distributions visually.
MTB > RANDOM 300 observations and put into C1;
SUBC> NORMAL MU=0 SIGMA=1.
MTB > #The following command squares each entry
MTB > #in C1 and puts it in C2.
MTB > LET C2= (C1)**2
MTB > HISTOGRAM C2
```

```
Histogram of C2 N = 300
Each * represents 5 obs.

Midpoint  Count
       0    162 *********************************
       1     72 ***************
       2     32 *******
       3     13 ***
       4     10 **
       5      5 *
       6      4 *
       7      0
       8      1 *
       9      1 *

MTB > RANDOM 300 observations and put into C3;
SUBC> CHISQUARE V=1.
MTB > HISTOGRAM C3

Histogram of C3 N = 300
Each * represents 5 obs.

Midpoint  Count
       0    145 *****************************
       1     79 ****************
       2     35 *******
       3     19 ****
       4     10 **
       5      8 **
       6      1 *
       7      1 *
       8      0
       9      0
      10      2 *

MTB > DESCRIBE C2 C3

          N     MEAN   MEDIAN   TRMEAN    STDEV   SEMEAN
C2      300   0.9743   0.4487   0.7740   1.4063   0.0812
C3      300   1.0938   0.5176   0.9021   1.4698   0.0849

        MIN      MAX       Q1       Q3
C2 0.0000   9.1355   0.0779   1.1768
C3 0.0001   9.8755   0.1028   1.5170
```

We note from the statistical description that the mean of both sampling distributions is about 1 and the standard deviation about 1.4. If we had used more than 300 observations to simulate the distributions these values would have approached closer to 1 and 1.414 respectively because for a χ^2 distribution with V degrees of freedom the mean is V, and the standard deviation is $\sqrt{2V}$. ($\sqrt{2 \times 1} = 1.414$).

```
MTB > #This simulates a chi-square distribution
MTB > #with 4 degrees of freedom from samples from
MTB > #a normal distribution.

MTB > BASE 1000
MTB > RANDOM 300 observations and put into C1-C4;
SUBC> NORMAL MU=0 SIGMA=1.
MTB > #This time instead of squaring each column
MTB > #using the LET command we will use the Row
MTB > #sum of squares RSSQ, which squares each
MTB > #column.
MTB > RSSQ C1-C4 and put it into C5
MTB > HISTOGRAM C5

Histogram of C5 N = 300
Each * represents 5 obs.

Midpoint    Count
       0       30 ******
       2      107 **********************
       4       84 *****************
       6       48 **********
       8       14 ***
      10       11 ***
      12        2 *
      14        2 *
      16        2 *
```

As can be seen, given the limitation of Minitab graphics, the above χ^2 distribution for 4 degrees of freedom corresponds to figure 11.1.

Postscript

In this chapter we looked briefly at the χ^2 distribution. We found that, as with the normal distribution from which it is derived, the χ^2 is really a family of distributions with an infinite number of members. Different members of the family are distinguished in shape by their degree of freedom. The change in shape of the distribution for low degrees of freedom, say less than eight, is very marked.

The χ^2 was used in a goodness-of-fit table to examine how well observed data matched a hypothetical model. We compared tables of observed values categorized in two dimensions (contingency tables) with expected values based on an equal expectation hypothesis. The fit between the observed and expected values was measured by squaring and then summing the 'average' differences in each cell. We ended the chapter by simulating sampling distributions for various degrees of freedom.

Exercises

1 A die was rolled 60 times and the following results were observed: 6 ones, 8 twos, 16 threes, 17 fours, 8 fives and 5 sixes. Make a decision at the 0.05 significance level whether the die is fair or not.

2 In a study of bizarre episodes in two mental hospitals, the results in table 11.3 were recorded. Can it be concluded from this study that the timing of the episodes and the type of hospital are independent? (Use level of significance $\alpha = 0.05$.)

Table 11.3

	Time of Episode	
	Weekday	*Evening or weekends*
Private hospital	34	17
State hospital	27	19

3 A 100 randomly selected workers of various class origins in a London hospital stated their voting intentions as shown in table 11.4. Use Minitab to perform a χ^2 analysis and examine the null hypothesis that social class makes no difference to voting intentions. (Use level of significance $\alpha = 0.05$.)

Table 11.4

	Voting intention		
Social class origins	*Labour*	*Liberal/SDP*	*Conservative*
Working class	20	10	10
Middle class	10	20	30

4 Use Minitab to simulate a χ^2 sampling distribution with 8 degrees of freedom using 300 random samples from a standard normal distribution (start the random number generator at 1000, i.e. BASE 1000). Using the same starting base simulate 300 observations directly from the χ^2 distribution with 8 degrees of freedom. Plot the sampling distributions as histograms. Get a statistical description of the two distributions and check that their means and standard deviations are approximately equal to the number of degrees of freedom and $\sqrt{2}$ times the number of degrees of freedom respectively.

Answers to Exercises

1

1 Hypothesis to be tested: H_0: the die is fair. H_1: the die is not fair

2 Test statistic: The test statistic will be

$$\text{calc.}\chi^2 = \sum \frac{(o-e)^2}{e}$$

3 Significance level: $\alpha = 0.05$ ($P = 0.95$).

4 *Sampling distribution and critical value:* tab. χ^2 with $n-1 = 6-1 = 5$ df. Critical value 11.07 (Table 4 Appendix 2)

5 *Decision rule with $\alpha = 0.05$ and df = 5:*

reject H_0 if calc. $\chi^2 \geqslant 11.07$
accept H_0 if calc. $\chi^2 < 11.07$

6 *Value of calc. χ^2 on the assumption that H_0 is true:*

Outcome (die face)	1	2	3	4	5	6
Observed o	6	8	16	17	8	5
Expected e	10	10	10	10	10	10

$$\text{calc.}\chi^2 = \sum \frac{(o-e)^2}{e}$$

$$= \frac{(6-10)^2}{10} + \ldots + \frac{(5-10)^2}{10} = 13.4$$

7 *Decision:* The test statistic calc. χ^2 is 13.4; this is greater than the critical value tab. χ^2 of 11.07 ($\alpha = 0.05$, df = 5) and therefore by our decision rule (5) we must reject the null hypothesis H_0 that the die is fair. On this evidence the die seems to be significantly biased.

2

1 *Hypothesis to be tested:* H_0: the two variables, type of hospital and timing of bizarre episodes are independent. H_1: the two variables, type of hospital and timing of bizarre episodes are not independent.

2 *Test statistic:*

$$\text{calc.}\chi^2 = \sum \frac{(o-e)^2}{e}$$

3 *Significance level:* $\alpha = 0.05$.

4 *Sampling distribution and critical values:* tab. χ^2 with $(n-1)(m-1) = (2-1)(2-1) = 1$ df. Critical value 3.841 (Table 4 Appendix 2)

5 *Decision rule with $\alpha = 0.05$, df = 1:*
reject H_0 if calc. $\chi^2 \geqslant 3.841$
accept H_0 if calc. $\chi^2 < 3.841$

6 *Value of calc. χ^2 on assumption H_0 is true:*

Observed o				Expected e (rounded)		
	34	17	51		32	19
	27	19	46		29	17
	61	36	97			

$$\text{calc. } \chi^2 = \sum \frac{(o-e)^2}{e}$$

$$= \frac{(34-32)^2}{32} + \ldots + \frac{(19-17)^2}{17} = 0.709$$

7 *Decision:* The test statistic calc. χ^2 is 0.709; this is less than the critical value tab. χ^2 of 3.841 and therefore by our decision rule (5) we cannot reject H_0. It still stands. The two variables of time of bizarre episodes and type of hospital do seem to be independent in this study.

3

1 *Hypothesis to be tested:* H_0: the two variables of social class origin and stated voting intention are independent. H_1: the two variables of social class origin and stated voting intention are not independent.

2 *Test-statistic:*

$$\text{calc.} \chi^2 = \sum \frac{(o-e)^2}{e}$$

3 *Significance level:* $\alpha = 0.05$.

4 *Sampling distribution and critical value:* tab. χ^2 with $(n-1)(m-1) = (2-1)(3-1) = 1 \times 2 = 2$ df. Critical value 5.991 (Table 4 Appendix 2)

5 *Decision rule with $\alpha = 0.05$, df. = 2:*

reject H_0 if calc. $\chi^2 \geqslant 5.991$
accept H_0 if calc. $\chi^2 < 5.991$

6 *Value of calc. χ^2 on the assumption that H_0 is true:*

```
MTB > READ these data into C1-C3
DATA> 20 10 10
DATA> 10 20 30
DATA> END
      2 ROWS READ
MTB > CHISQUARE test on data in C1-C3

Expected counts are printed below observed counts

            C1       C2       C3     Total
1           20       10       10        40
         12.00    12.00    16.00

2           10       20       30        60
         18.00    18.00    24.00

Total       30       30       40       100

ChiSq =  5.333 + 0.333 + 2.250 +
         3.556 + 0.222 + 1.500 = 13.194
df = 2
```

7 *Decision:* The test statistic calc. χ^2 is 13.19; this is greater than the critical value tab. χ^2 of 5.991 and therefore by our decision rule (5) we must reject H_0 On the basis of this sample there is significant evidence that the two variables are not independent. Social class origin does appear to affect voting intentions. (From Neave it can be seen that if we had adopted the more stringent level of significance of $\alpha = 0.01$ and so had a tab. χ^2 of 9.21 our calc. χ^2 of 13.19 would still have been significant.)

4

```
MTB > #This is for 8 degrees of freedom first from
MTB > #squared normal distributions and then from
MTB > #samples using the chi-square distribution
MTB > #directly.
MTB > BASE 1000
MTB > RANDOM 300 observations and put into C1-C8;
SUBC> NORMAL MU=0 SIGMA=1.
MTB > #Remember the command RSSQ is a row command:
MTB > #it squares the values in each column and
MTB > #then sums these squares and puts them in
MTB > #the specified column.

MTB > RSSQ C1-C8 and put into C9
MTB > HISTOGRAM C9

Histogram of C9 N = 300
Each * represents 2 obs.

Midpoint Count
       0     1 *
       2    21 ***********
       4    41 *********************
       6    68 **********************************
       8    57 *****************************
      10    42 *********************
      12    25 *************
      14    18 *********
      16    14 *******
      18     6 ***
      20     3 **
      22     1 *
      24     2 *
      26     1 *


MTB > #Directly from a chi-square distribution.
MTB > BASE 1000
MTB > RANDOM 300 observations and put into C10;
SUBC> CHISQUARE V=8.
MTB > HISTOGRAM C10
```

```
Histogram of C10 N = 300
Each * represents 2 obs.

Midpoint   Count
     0         1 *
     2        26 *************
     4        62 *******************************
     6        61 *******************************
     8        47 ************************
    10        38 *******************
    12        31 ****************
    14        17 *********
    16        12 ******
    18         1 *
    20         2 *
    22         2 *
```

```
MTB > DESCRIBE C9 C10
         N      MEAN   MEDIAN   TRMEAN   STDEV   SEMEAN
C9     300     8.439    7.742    8.173   4.316    0.249
C10    300     7.748    7.002    7.537   4.044    0.233

       MIN       MAX       Q1        Q3
C9   0.837    25.585    5.336    10.571
C10  0.793    22.862    4.737    10.684
```

As can be seen both of the above give similar histograms (sampling distributions) and their means and standard deviations are as follows:

C9 mean, 8.439 standard deviation, 4.316
C10 mean, 7.748 standard deviation, 4.044

Both means are close to 8, the number of degrees of freedom, and the standard deviations are not far from the theoretical value of the square root of twice the number of degrees of freedom, i.e. $\sqrt{2 \times 8} = \sqrt{16} = 4$.

Afterword

We have now come to the end of this introductory text. If you have worked through all the examples in the preceding chapters for yourself and have done some of the exercises in each, you should have a good grasp of the basic concepts of elementary data analysis.

You should be in a position to use the tools that we have developed on data of your own choosing or for course project work. What you have learnt should also provide a good basis upon which to develop your skills in data analysis should you so wish. It is a fascinating field of study and one which is not only expanding rapidly but also becoming more accessible with the development of cheap and powerful personal computers.

As this is an introductory text there are of course areas of data analysis that I have omitted and also much that I have not proved rigorously but merely stated and in some cases also demonstrated. The results deriving from the central limit theorem which link the population parameters of mean and standard deviation to the sample statistics are in this category. You will remember that I asked you to take the results on trust, but we did demonstrate their validity by computer simulation. Similarly, elsewhere I have quoted rather than derived various formulae. You should also note that we have not covered such important areas of data analysis as the collection of data, sampling and the design of simple studies. We did emphasize, however, the importance of random sampling and its ability to produce probability distributions in chapters 9 and 10.

Should you now wish to practise the skills you have learnt on easily accessible data sets I suggest that you focus initially on the abstract of UNICEF data in the file DATA11.DAT and examine, say, the GNP and some of the variables that we have not used. I have also included another suicide data file, DATA01.DAT, that you may wish to analyse and compare with the results of the work that we did in chapter 5, or indeed with the earlier analysis of similar data by Erikson and Nosanchuk (1992). The Minitab package itself has a collection of 31 data files built into it. These cover a variety of data from various areas of science, social science etc. to the letters of Mark Twain and the blood pressure of Peruvian Indians. A list and a brief description of these files is to be found in the *Minitab Handbook* by Ryan, Joiner and Ryan.

To call these files into Minitab you should first ask for the directory of the files with the command MTB > DIR C:\MINITAB\DATA. This will display the set of resident files. To retrieve them into the data editor you will need to use commands such as

```
MTB > RETRIEVE 'C:\MINITAB\DATA\PERU.MTW'
```

or

```
MTB > RETRIEVE 'C:\MINITAB\DATA\PULSE.MTW'
```

etc. For the documentation relating to various files you will need to consult the *Minitab Handbook*.

If you do work on the above data files from Minitab or the ones that I have provided or indeed on course project work, you will be faced with a problem you have not encountered in working through this text. The problem is knowing where to start your analysis and in what direction to pursue it.

Let me end this Afterword by offering a few hints in tackling the elementary analysis of a data set on your own.

The analysis exercises that we have done in this book have had a necessary artificiality about them in that they have all been chapter linked. That is, the worked examples in chapter Q concern the subject studied in chapter Q. The choice of where to start and the type of analysis to employ has already been made for you. However, when you are analysing your own data or a data file such as those mentioned above it is not so easy to decide where to start the analysis and what tools to use.

By way of illustration let us refer back to the abstract of UNICEF statistics in DATA11.DAT. Let us imagine for a moment that we have done no work on this data set but that we are facing it for the first time. We must decide what general direction our analysis is to take and how to start going about it. We need an overview or general goal and we need to decide on the means to achieve it. In short we need a long-term strategy and intermediate tactics.

Let me mention one place that you should *not* start. In front of the keyboard! It should be stressed that neither the strategy nor the tactics can be adequately worked out by your sitting in front of the data on the screen, with hands poised above the keyboard and itching to 'get on with it'.

The initial thinking should be done away from the computer. You first need to develop a flexible overall strategy. Let us say that we have decided to start analysing DATA11.DAT with the overall aim of looking at the relationship between the larger variables of wealth and health (strategy). Looking at a copy of the data (see appendix 1, p.366–9) we see that there is one major index of wealth in the data, the GNP per capita, and three indices of health: infant mortality, maternal mortality and life expectancy.

In order to plan the initial part of our analysis (tactic) we can choose to look first at the one index of wealth, GNP, together with one index of health, IMR. It is worth drafting a few ideas with pencil and paper *before* going to the computer. These could be rough handwritten notes on the following lines. The hash signs # indicate suitable places for comment etc.

```
> >OW=70 OUTFILE 'A:\BLODWEN1'
> #Overall strategy. To link wealth and health.
> #Using various indices.
> #First look at GNP and IMR.
> READ 'A:\DATA11.DAT' into C1-C12
> #Name relevant columns.
> PLOT C7 against C2
> #Comment ... strong pattern in data etc. ...
> #OK, but which countries are which on plot?
```

```
> #LPLOT by U5MR and individually.
> LPLOT C7 against C2 using tags in C1
> # Comment ... set up column for individual tags.
> SET C13
DATA> 1:59
DATA> END
> LPLOT C7 against C2 using tags in C13
> #Comment ... odd values? outliers?
> #Look now at GNP by itself, its distribution etc.
> #Use DESCRIBE, BOXPLOT, HISTOGRAM, SCATTER of C2
> #Comment ...
> #DESCRIBE, BOXPLOT etc. by U5MR group in C1

> DESCRIBE C2;
SUBC> BY C1.
> BOXPLOT C2;
SUBC> BY C1. etc. etc.
> #Comment ...
> #Do similar types of analysis of IMR on its own.
> #Is there time today?
> PRINT C1-C? #To print the data into the
> #analysis file.
> NOOUTFILE
> SAVE 'A:\BLODWEN1' #If data have been
> #substantially modified.
> STOP
> #Print out the analysis A:\BLODWEN1.LIS and
> #take it home for further
> #close study.
```

With the above rough 'hand-written' guide go to the computer and complete this part of the analysis, remembering to comment on each plot or presentation, no matter how briefly. At the end of the session print out the analysis file and study it carefully. Later the file BLODWEN1.LIS can be edited on a word-processing package to form a part of the completed analysis.

Study the print-out and think what you have found in the light of your overall strategy of relating wealth and health. If what you have done seems to fit this strategy you could then write a few further notes about this first leg of the analysis and think what you wish to do at the next session. You may for instance decide to plot GNP against IMR again, straighten out the data with a suitable transformation and fit a regression line. Again it is worth thinking about this *away* from the computer and drafting out guidelines and commands with pen and paper similar to the above.

If the line fitting is successful it may be time to look at the GNP and some of the other indices of health – MMR and life expectancy. Plan and carry this out in a similar orderly fashion, remembering always to comment on your analyses as you proceed and of course being careful to save data and analyses in a neat and orderly way for easy retrieval and editing.

At this stage you need to ask how the analyses completed so far inform or change your view of the overall relationship between the larger variables of wealth and health. Is it useful to bring in other factors at this stage? What about education? What indices of this do we have in the data set? Remembering what we found earlier regarding female literacy, education is probably an important factor to consider.

So, to summarize, for a successful analysis we need an overall goal or strategy and a series of tactical planned sessions on the computer to provide the means of achieving this goal. The end product of the analysis should of course be a written report detailing the findings and conclusions and giving a critical summary of both the analysis and the data set etc. With this concluding report you should provide an edited copy of selected parts of your progressive analysis. Finally you should write a brief introduction to signpost and help the reader through the work.

Appendix 1

This appendix contains eleven ASCII data files that can be read by Minitab or a word-processor package. The commands to read the files into Minitab are given with each file.

```
#This is file DATA01.DAT
#Call into Minitab with MTB>READ 'A:\DATA01.DAT' into C1-C9.
#Age specific death rates per 100,000 population for suicide.
#World Health Statistics Annual, WHO, Geneva, 1991 & 1992.
#1= Male, 0= Female. Age bands in years. All indicates overall rate.
#Ctry Sex All     15-24   25-34   35-44  45-54  55-64   65-74  75+
#
-Aust    1 34.8    25.0    32.7    37.3   47.7   44.0    66.8   107.6
         0 13.4    5.5     8.4     13.9   16.5   20.8    22.0   35.5
#
-Fran    1 30.5    15.8    32.2    37.4   39.2   38.9    52.9   107.2
         0 11.7    4.6     9.2     13.0   17.4   19.4    20.2   25.7
#
-WGer    1 23.5    14.7    22.7    21.6   30.8   31.5    39.4   75.6
         0 10.0    4.2     6.9     7.9    13.0   13.7    18.2   22.7
#
-Hung    1 59.9    20.1    56.0    79.5   99.6   90.2    97.3   196.6
         0 21.4    8.2     11.8    20.3   29.9   28.8    37.1   75.6
#
-Irel    1 12.1    13.3    17.3    20.5   17.9   14.8    20.0   18.1
         0 3.7     2.4     3.3     4.5    6.9    8.6     9.2    5.3
#
-Isra    1 9.7     9.1     12.7    11.3   18.6   18.3    23.4   30.6
         0 4.0     2.1     2.5     6.4    6.3    10.1    10.3   15.9
#
-Ital    1 11.7    5.1     10.3    10.0   14.1   18.5    29.1   47.1
         0 4.2     1.4     2.4     3.6    5.6    7.5     9.0    11.7
#
-Neth    1 13.0    8.9     15.4    15.6   17.3   19.9    21.1   29.0
         0 7.5     4.4     7.2     7.5    13.8   12.6    13.4   10.8
#
-Pola    1 22.0    16.2    30.4    33.0   38.4   36.3    28.0   27.3
         0 4.5     2.8     4.3     6.0    7.4    8.7     7.2    7.4
#
-Spai    1 10.6    6.4     9.8     8.5    13.3   18.7    25.7   45.5
         0 4.0     2.1     3.2     3.3    4.7    7.3     9.8    10.1
#
-Swed    1 26.4    18.2    29.8    31.8   38.2   37.0    36.4   47.2
         0 11.5    5.9     11.3    17.1   20.5   14.5    15.3   13.8
#
-Swit    1 31.5    24.8    33.2    32.2   39.0   42.7    51.6   86.8
         0 12.7    6.3     10.9    12.8   19.6   16.3    23.1   23.5
#
-E&W     1 12.1    11.7    16.0    17.1   16.4   13.3    13.6   19.4
         0 3.7     2.0     3.8     4.5    5.1    5.3     6.1    6.2
#
-Cana    1 20.9    26.4    28.3    25.8   24.5   26.0    25.3   29.5
         0 6.0     4.7     7.3     8.9    10.4   7.9     6.9    5.4
#
-USA     1 20.1    21.9    25.0    22.9   21.7   25.0    33.0   57.8
         0 5.0     4.2     5.7     6.9    7.9    7.2     6.8    6.4
#
-Japa    1 20.4    9.2     18.4    21.5   32.0   32.5    36.6   62.9
         0 12.4    4.7     9.0     9.2    15.0   17.6    25.3   48.6
```

```
#This is the file DATA02.DAT.
#Call into Minitab with MTB>READ 'A:\DATA02.DAT' into C1-C3
#It contains the heights of 100 men in C2 and 100 women in C3.

  1    173    155
  2    181    169
  3    169    162
  4    183    166
  5    179    176
  6    167    162
  7    183    160
  8    172    144
  9    182    159
 10    175    159
 11    176    162
 12    173    149
 13    182    167
 14    173    162
 15    170    161
 16    173    163
 17    172    153
 18    176    163
 19    180    153
 20    177    143
 21    186    164
 22    168    156
 23    158    159
 24    175    166
 25    174    162
 26    177    158
 27    170    150
 28    179    166
 29    192    160
 30    164    169
 31    175    168
 32    179    148
 33    177    170
 34    177    168
 35    173    172
 36    162    156
 37    160    160
 38    169    161
 39    167    157
 40    175    158
 41    182    158
 42    166    165
 43    179    170
 44    174    163
 45    165    165
 46    176    162
 47    174    156
 48    175    155
 49    178    157
 50    174    164
 51    179    167
 52    172    164
 53    193    154
 54    171    155
 55    186    161
 56    180    154
 57    173    157
 58    170    160
 59    176    162
 60    181    169
 61    172    163
 62    170    165
 63    174    158
 64    171    164
 65    171    169
 66    157    161
 67    174    163
 68    170    166
 69    180    153
 70    162    170
 71    181    156
 72    172    163
 73    169    161
 74    175    156
 75    178    150
 76    163    171
 77    172    160
 78    181    151
 79    188    163
 80    188    167
 81    175    165
 82    173    174
 83    169    170
 84    166    159
 85    174    167
 86    173    165
 87    174    168
 88    165    150
 89    166    172
 90    171    157
 91    177    165
 92    184    160
 93    168    154
 94    169    164
 95    166    158
 96    177    160
 97    168    170
 98    175    160
 99    181    164
100    175    159
#
```

```
#This is file DATA03.DAT.
#Call into Minitab with MTB>READ 'A:\DATA03.DAT' into C1-C3.
#It contains the heights & weights of 45 male & 39 female
#undergraduates      (1989).
#Heights in cms. to the nearest cm. in C2.
#Weights in kgs. to the nearest kg. in C3.
#C1 sex Male=1 Female=0   * indicates a missing value.
#
1    158  63
1    159  59
1    164  64
1    166  62
1    166  61
1    167  57
1    169  67
1    170  66
1    170  70
1    171  74
1    171  69
1    172  70
1    173  73
1    174  67
1    174  73
1    175  *
1    175  66
1    176  67
1    176  68
1    176  74
1    176  78
1    177  70
1    177  68
1    177  79
1    178  73
1    178  *
1    178  76
1    178  60
1    179  68
1    180  80
1    180  77
1    180  71
1    181  69
1    182  82
1    182  83
1    183  77
1    185  82
1    185  80
1    186  77
1    188  78
1    189  85
1    190  *
1    194  80
1    196  73
1    199  76
0    148  50
0    150  52
0    151  47
0    154  49
0    154  48
0    155  *
0    155  56
0    156  *
0    157  53
0    157  50
0    157  *
0    158  56
0    159  52
0    159  53
0    160  55
0    160  *
0    161  59
0    161  58
0    161  55
0    162  55
0    163  59
0    164  56
0    164  53
0    164  57
0    164  62
0    164  56
0    165  57
0    165  60
0    165  61
0    166  55
0    167  61
0    168  66
0    170  64
0    173  68
0    174  57
0    174  71
0    181  70
0    182  65
0    185  67
#
```

```
# This is DATA04.DAT.
# Call into Minitab with MTB>READ 'A:\DATA04.DAT' into C1-C9.
# Age specific death rates per 100,000 population for suicide.
# World Health Statistics Annual 1988 .WHO Geneva.
# 1 = Male 0 = Female.Age bands in years.All indicates all ages.
# Country Sex All  15-24 25-34 35-44 45-54 55-64 65-74  75 +
#         c1  c2    c3    c4    c5    c6    c7    c8    c9
#
-Austria  1  40.1  29.3  42.9  44.8  54.9  48.5  68.1  125.2
          0  15.7  8.1   10.7  15.9  16.8  22.9  33.8  35.4
#
-France   1  32.9  16.0  34.2  38.1  46.3  47.8  63.5  121.3
          0  12.9  4.6   10.1  14.4  20.3  21.2  25.4  29.5
#
-W.German 1  26.7  17.6  24.3  27.7  35.2  37.5  43.9  77.2
          0  11.8  4.5   7.8   10.4  15.6  17.4  23.2  23.7
#
-Hungary  1  65.9  24.3  67.3  89.3  111.5 96.5  118.4 188.0
          0  25.6  10.3  17.9  25.8  33.7  37.3  56.4  69.0
#
-Ireland  1  12.1  14.0  18.7  13.1  17.7  28.3  20.5  8.8
          0  3.8   3.3   5.2   5.3   9.2   8.3   4.7   1.1
#
-Israel   1  8.5   6.6   7.7   9.8   13.9  19.7  26.4  39.8
          0  4.5   4.5   3.7   5.6   4.6   12.6  8.8   17.5
#
-Italy    1  12.2  5.2   9.2   10.1  14.8  22.0  33.7  50.0
          0  4.7   1.3   2.7   4.3   5.9   8.8   11.8  12.4
#
-Nlands   1  13.9  8.1   15.2  15.3  19.6  22.6  25.5  42.8
          0  8.2   3.6   9.1   10.9  13.6  13.5  12.5  13.0
#
-Poland   1  22.3  17.8  30.6  32.3  38.5  35.0  29.9  33.2
          0  4.7   3.0   5.7   6.4   8.0   7.7   7.7   6.9
#
-Spain    1  9.8   5.9   7.9   9.3   15.1  19.0  22.2  38.4
          0  3.4   1.6   2.2   2.5   4.8   6.5   7.9   10.9
#
-Sweden   1  27.1  19.5  29.4  33.8  38.8  36.1  39.3  48.0
          0  10.1  7.9   11.0  14.6  15.4  15.3  12.7  8.8
#
-Switzerl 1  35.0  27.7  37.3  41.3  44.5  43.3  56.7  87.3
          0  13.7  8.2   11.8  13.8  15.1  18.7  30.0  27.4
#
-Eng&Wals 1  11.6  9.3   14.6  14.6  15.0  17.8  15.8  20.0
          0  4.5   2.1   3.8   4.6   7.6   7.3   8.1   6.9
#
-Canada   1  22.8  26.9  32.0  28.5  28.1  27.6  28.4  36.3
          0  6.4   5.3   7.5   8.9   11.4  8.9   9.2   5.5
#
-USA      1  20.6  21.7  25.5  23.0  24.4  26.7  35.5  56.0
          0  5.4   4.4   5.9   7.6   8.8   8.4   7.3   6.8
#
-Japan    1  25.6  11.6  23.7  29.4  45.5  40.5  42.1  73.0
          0  13.8  6.5   9.9   10.9  16.9  21.2  31.1  53.2
#
```

```
# This is datafile DATA05.DAT
# Call into Minitab with MTB>READ 'A:\DATA05.DAT' into C1-C2.
# This is a file of the heights & weights of 60 seamen.
# Weights to the nearest kg.(C1) Heights to the nearest cm.(C2)
#  Name          Weight     Height
-Baker           85         188
-Beard           69         173
-Breedon         75         183
-Brookes         58         181
-Burman          66         178
-Cameron         48         165
-Cheeseman       73         173
-Coleman         75         191
-Coombe          62         177
-Day             73         185
-Doughty         46         163
-Dunkley         73         178
-Eldridge        58         175
-Elworthy        73         181
-Ewan            75         188
-Fox             71         175
-Gaukroger       64         183
-Gibbons         67         173
-Goodsir         59         178
-Griffiths       55         172
-Hardy           61         180
-Harper          70         173
-Hart            54         179
-Hawkins         68         178
-Hicks           75         187
-Hill            57         168
-Holloway        77         178
-Horsfield       68         175
-Isherwood       53         168
-Jarvis          62         178
-Joshua          74         185
-McDonald        65         178
-Maidment        69         170
-Maund           60         173
-Mawford         50         175
-Middleton       61         173
-Mitchell        80         188
-Morgan          59         179
-Munro           67         175
-Newton          64         176
-Parr            67         179
-Parry           62         177
-Pegram          80         180
-Percy           71         185
-Phipps          51         170
-Pizzey          55         175
-Rentell         62         183
-Ross            51         169
-Ruegg           48         173
-Sargent         68         180
-Smith           63         176
-Stanness        57         170
-Stares          71         173
-Straford        60         170
-Teagles         53         169
-Tearle          68         180
-Wells           54         171
-Woodgate        67         179
-Youngman        71         182
-Zambelli        63         176
#
```

```
#This is data file DATA06.DAT
#Call into Minitab with MTB>READ 'A:\DATA06.DAT' into C1-C2.
#Incidence of low birthweight by height of mother,Scotland 1979.
#Birth Counts,Alison Macfarlane,Miranda Mugford, 1984.
#X maternal height in cms.(C1)
#Y % of birthweight less than 2500 gms.(C2)
# X       Y
146      16.0
153      10.0
158      8.3
163      5.8
168      4.0
172      3.3
180      1.9
#
```

```
#This is data file DATA07.DAT
#Call into Minitab with MTB>READ 'A:\DATA07.DAT' into C1-C2.
#Abortions in thousands (residents only) England & Wales.
#Source, Population Trends,69,Autumn 1992, OPCS.
#Year     No.of women(Thousands)
1971       94.6
1976      101.9
1981      128.6
1983      127.4
1984      136.4
1985      141.1
1986      147.6
1987      156.2
1988      168.3
1989      170.5
1990      173.9
1991      167.4
#
```

```
#This is data file DATA08.DAT
#Call into Minitab with MTB>READ 'A:\DATA08.DAT' into C1-C3.
#Paul T Young,Motivation & Emotion,1961. John Wiley.  p.450
#Development of retaliative behavior in response to frustration.
#
#Mean age   % outbursts     % outbursts
#in years   undirected      directed
#           energy.         retaliatory behaviour
#
0.5          88.9            0.7
1.5          78.4            6.3
2.5          75.1           10.6
3.5          59.9           25.6
6.0          36.3           28.0
#
#2124 observed responses on 45 male & female
#children by Goodenough F.L.
#
```

```
#This is a data file DATA09.DAT
#Call into Minitab with MTB>READ 'A:\DATA09.DAT' into C1-C2.
#The population of Ireland from Mitchell B.R., European Historical
#Statistics,1981.
#Year    Population in thousands.
1821      6802
1831      7767
1841      8175
1851      6552
1861      5799
1871      5412
1881      5175
1891      4705
1901      4459
1911      4390
#
```

```
#This is file DATA10.DAT
#Call into Minitab with MTB>READ 'A:\DATA10.DAT' into C1-C2.
#Irish alcohol consumption per head of population
#aged 15 years and over,in pure litres of 100% alcohol 1950-1987.
#Source,Annual reports of Revenue Commissions, Irish Statistical
#Bulletin,1989.
#
#Year      Total consumption per capita
1950       4.67
1951       4.91
1952       4.41
1953       4.40
1954       4.49
1955       4.62
1956       4.68
1957       4.54
1958       4.49
1959       4.67
1960       4.80
1961       5.30
1962       5.24
1963       5.44
1964       5.71
1965       5.83
1966       5.80
1967       5.91
1968       6.27
1969       6.80
1970       7.13
1971       7.57
1972       8.09
1973       8.80
1974       9.44
1975       9.15
1976       8.85
1977       9.09
1978       9.64
1979       9.96
1980       10.90
1981       10.00
1982       9.50
1983       9.40
1984       9.60
1987       7.60
#
```

```
#This is the data file DATA11.DAT
#Call into Minitab with MTB>READ 'A:\DATA11.DAT' into C1-C12.
#The data is abstracted from various tables in James Grant,
#The State of the World's Children 1992,  United Nations
#Children's Fund,  Oxford University Press.
#      1     2      3     4     5     6     7     8     9    10    11     12
-Moza  1    80      5    10    35    17   173    45    21   300    48   15.7
-Ango  1   610      6    15    34     *   173    56    29     *    46   10.0
-Mala  1   180      7    11     7    17   144     *     *   170    48    8.8
-Ethi  1   120      *     *     *    34   130     *     *     *    46   49.2
-Nepa  1   180      6     9     7    14   123    38    13   830    52   19.1
-Bang  1   180     10    11    10    13   114    47    22   600    52  115
-Suda  1   420      *     *     *     5   104    43    12   550    51   25.2
-Tanz  1   130      6     8    16    13   102    93    88   340    54   27.3
-Nige  1   250      1     3     3    21   101    62    40   800    52  108
-Ugan  1   250      2    15    26    45    99    62    35   300    52   18.8
-Boli  1   620      1    12     6    26   102    85    71   600    55    7.3
-Paki  1   370      1     3    30    16   104    47    21   500    58  123
-Indi  1   340      2     3    19    19    94    62    34   460    59  853
#
-Ghan  2   390      9    26     3    22    86    70    51  1000    55   15.0
-Zamb  2   390      7     9     *    11    76    81    65   150    54    8.5
-Peru  2  1010      6    16    20     4    82    92    79   300    63   21.6
-Liby  2  5310      *     *     *     *    75    75    50    80    62    4.5
-Moro  2   880      3    17    15    26    75    61    38   300    62   25.1
-Keny  2   360      5    19     8    19    68    80    59   170    60   24.0
-Alge  2  2230      *     *     *    68    68    70    46   140    65   25.0
-Indo  2   500      2    10     8    27    71    84    62   450    61  184
-Saud  2  6020      *     *     *     *    65    73    48     *    65   14.1
-Safr  2  2470      *     *     *     *    67     *     *    83    62   35.3
-Zimb  2   650      *    23    16    20    61    74    60     *    60    9.7
-Iraq  2  2340      *     *     *     *    63    70    49   120    65   18.9
-Egyp  2   640      3    12    20    17    61    63    34   320    60   52.4
-Turk  2  1370      2    13    10    28    69    90    71   150    65   55.9
-Braz  2  2540      6     3     3    19    60    83    80   200    66  150
#
-Phil  3   710      4    17    13    20    43    90    90   100    64   62.4
-Viet  3   240      *     *     *     *    49    92    84   120    63   66.7
-Tuni  3  1260      6    15     6    21    48    74    56    50    67    8.2
-Iran  3  3200      6    20    14     *    46    65    43   120    66   54.6
-Jord  3  1640      4    14    30    15    40    89    70    48    67    4.0
-Colo  3  1200      *     *     *    38    39    88    86   200    69   33.0
-Mexi  3  2010      1     7     1    26    40    90    85   110    70   88.6
-Chin  3   350      *     *     8     8    30    84    62    95    70 1139
-Arge  3  2160      2     6     7    23    31    96    95   140    71   32.3
-Sril  3   430      6    11     7    13    26    93    84    80    71   17.2
-Thai  3  1220      6    19    19     8    26    96    90     *    66   55.7
-UAE   3 18430      7    15    44     *    24    58    38     *    70    1.6
-Maly  3  2160      5     *     *    12    22    87    70    59    70   17.9
-Chil  3  1770      4    12     *    15    20    94    93    67    72   13.2
#
-Jama  4  1260      7    11     8    16    16    98    99   120    73    2.5
-Hung  4  2590      2     2     5    23    15     *     *    15    71   10.6
-Cuba  4  1170     23    10     *     *    11    95    93    39    75   10.6
-Newz  4 12070     12    11     5     *    10     *     *    13    75    3.4
-Isra  4  9790      4    10    27     *    10     *     *     3    76    4.6
-USA   4 20910     13     2    25     *     9     *     *     8    76  249
-Ital  4 15120     10     8     3     *     9    98    96     4    76   57.1
-Aust  4 14360     10     7     9     *     8     *     *     3    77   16.9
```

#	1	2	3	4	5	6	7	8	9	10	11	12
-Spai	4	9330	13	6	6	*	8	97	93	5	77	39.2
-UK	4	14610	13	2	13	*	8	*	*	8	76	57.2
-Fran	4	17820	21	8	6	*	8	*	*	9	76	56.1
-Denm	4	20450	1	9	5	*	8	*	*	3	76	5.1
-Irel	4	8710	13	11	3	*	8	*	*	2	75	3.7
-Germ	4	20440	*	*	*	*	7	*	*	*	75	77.6
-Swed	4	21570	1	9	7	*	6	*	*	5	77	8.4
-Japa	4	23810	*	*	*	*	5	*	*	11	79	123
-Swiz	4	29880	13	3	10	*	7	*	*	5	77	6.6

#1 Mozambique
#2 Angola
#3 Malawi
#4 Ethiopia
#5 Nepal
#6 Bangladesh
#7 Sudan
#8 Tanzania
#9 Nigeria
#10 Uganda
#11 Bolivia
#12 Pakistan
#13 India
#14 Ghana
#15 Zambia
#16 Peru
#17 Libya
#18 Morocco
#19 Kenya
#20 Algeria
#21 Indonesia
#22 Saudi Arabia
#23 South Africa
#24 Zimbabwe
#25 Iraq
#26 Egypt
#27 Turkey
#28 Brazil
#29 Philippines
#30 Viet Nam
#31 Tunisia
#32 Iran
#33 Jordan
#34 Colombia
#35 Mexico
#36 China
#37 Argentina
#38 Sri Lanka
#39 Thailand
#40 United Arab Emirates
#41 Malaysia
#42 Chile
#43 Jamaica
#44 Hungary
#45 Cuba
#46 New Zealand
#47 Israel
#48 United States of America
#49 Italy
#50 Australia
#51 Spain
#52 United Kingdom
#53 France
#54 Denmark
#55 Ireland
#56 Germany
#57 Sweden
#58 Japan
#59 Switzerland

```
#    C1   Type of country U5MR Under five mortality rate: Annual
#         number of deaths of children under 5 years of age per
#         1000 live births
#         1    Very high U5MR - over 140
#         2    High U5MR      - 71-140
#         3    Middle U5MR    - 21-70
#         4    Low U5MR       - 20 and under
#
#    C2   Annual Gross National Product per capita in US dollars
#
#         % of Central Government Expenditure 1986-90, allocated
#         to:
#    C3   Health
#    C4   Education
#    C5   Defence
#
#    C6   Debt service % of Export of Goods and Services 1989
#
#    C7   Infant Mortality Rate IMR.  Annual number of death of
#         infants under 1 year of age per 1000 live births
#         1990.
#
#    C8   Adult male literacy rate 1990 ie % of males aged 15
#         years and over who can read and write
#    C9   Adult female literacy rate 1990 ie % of females ages
#         15 years and over who can read and write
#
#    C10  Maternal Mortality Rate 1989-90.  Annual number of
#         death of women from pregnancy related causes per
#         100,000 live births
#
#    C11  Life expectancy at birth 1990 in years
#
#    C12  Total population in millions 1990
```

Appendix 2

This appendix contains selected tables from Henry Robert Neave, *Elementary Statistical Tables*, 1992 Routledge and Kegan Paul.

Table 1 Probabilities and ordinates in the normal distribution
Table 2 Percentage points of the normal distribution
Table 3 Percentage points of the student t distribution
Table 4 Percentage points of the Chi-squared distribution
Table 5 Critical values for Spearman's rank correlation coefficient

Table 1 Probabilities and ordinates in the normal distribution

$$\phi(z) = \frac{1}{\sqrt{2\pi}} e^{-\frac{1}{2}z^2}; \quad \Phi(z) = \text{prob}(Z) = \int_{-\infty}^{z} \phi(t)\, dt$$

$\phi(z)$	z	0	1	2	3	4	5	6	7	8	9
0.00443	−3.0	0.00135	00131	00126	00122	00118	00114	00111	00107	00104	00100
0.00595	−2.9	0.00187	00181	00175	00169	00164	00159	00154	00149	00144	00139
0.00792	−2.8	0.00256	00248	00240	00233	00226	00219	00212	00205	00199	00193
0.0104	−2.7	0.00347	00336	00326	00317	00307	00298	00289	00280	00272	00264
0.0136	−2.6	0.00466	00453	00440	00427	00415	00402	00391	00379	00368	00357
0.0175	−2.5	0.00621	00604	00587	00570	00554	00539	00523	00508	00494	00480
0.0224	−2.4	0.00820	00798	00776	00755	00734	00714	00695	00676	00657	00639
0.0283	−2.3	0.0107	0104	0102	0099	0096	0094	0091	0089	0087	0084
0.0355	−2.2	0.0139	0136	0132	0129	0125	0122	0119	0116	0113	0110
0.0440	−2.1	0.0179	0174	0170	0166	0162	0158	0154	0150	0146	0143
0.0540	−2.0	0.0228	0222	0217	0212	0207	0202	0197	0192	0188	0183
0.0656	−1.9	0.0287	0281	0274	0268	0262	0256	0250	0244	0239	0233
0.0790	−1.8	0.0359	0351	0344	0336	0329	0322	0314	0307	0301	0294
0.0940	−1.7	0.0446	0436	0427	0418	0409	0401	0392	0384	0375	0367
0.1109	−1.6	0.0548	0537	0526	0516	0505	0495	0485	0475	0465	0455
0.1295	−1.5	0.0668	0655	0643	0630	0618	0606	0594	0582	0571	0559
0.1497	−1.4	0.0808	0793	0778	0764	0749	0735	0721	0708	0694	0681
0.1714	−1.3	0.0968	0951	0934	0918	0901	0885	0869	0853	0838	0823
0.1942	−1.2	0.1151	1131	1112	1093	1075	1056	1038	1020	1003	0985
0.2179	−1.1	0.1357	1335	1314	1292	1271	1251	1230	1210	1190	1170
0.2420	−1.0	0.1587	1562	1539	1515	1492	1469	1446	1423	1401	1379
0.2661	−0.9	0.1841	1814	1788	1762	1736	1711	1685	1660	1635	1611
0.2897	−0.8	0.2119	2090	2061	2033	2005	1977	1949	1922	1894	1867
0.3123	−0.7	0.2420	2389	2358	2327	2296	2266	2236	2206	2177	2148
0.3332	−0.6	0.2743	2709	2676	2643	2611	2578	2546	2514	2483	2451
0.3521	−0.5	0.3085	3050	3015	2981	2946	2912	2877	2843	2810	2776
0.3683	−0.4	0.3446	3409	3372	3336	3300	3264	3228	3192	3156	3121
0.3814	−0.3	0.3821	3783	3745	3707	3669	3632	3594	3557	3520	3483
0.3910	−0.2	0.4207	4168	4129	4090	4052	4013	3974	3936	3897	3859
0.3970	−0.1	0.4602	4562	4522	4483	4443	4404	4364	4325	4286	4247
0.3989	−0.0	0.5000	4960	4920	4880	4840	4801	4761	4721	4681	4641
$\phi(z)$	z	0	1	2	3	4	5	6	7	8	9

The left-hand column gives the ordinate $\phi(z) = e^{-1/2z^2}/\sqrt{(2\pi)}$ of the standard normal distribution (i.e. the normal distribution having mean 0 and standard deviation 1), z being listed in the second column. The rest of the table gives $\Phi(z) = \int_{-\infty}^{z} \phi(t)dt = \text{prob}(Z \leqslant z)$, where Z is a random variable having the standard normal distribution. Locate z, expressed to its first decimal place in the second column, and its second decimal place along the top or bottom horizontal: the corresponding table entry is $\Phi(z)$.

EXAMPLES: $\Phi(-1.2) = \text{prob}(Z \leqslant -1.2) = 0.1151$;
$\Phi(-1.23) = 0.1093$; $\Phi(\theta\ 1.234) = 0.1086$.

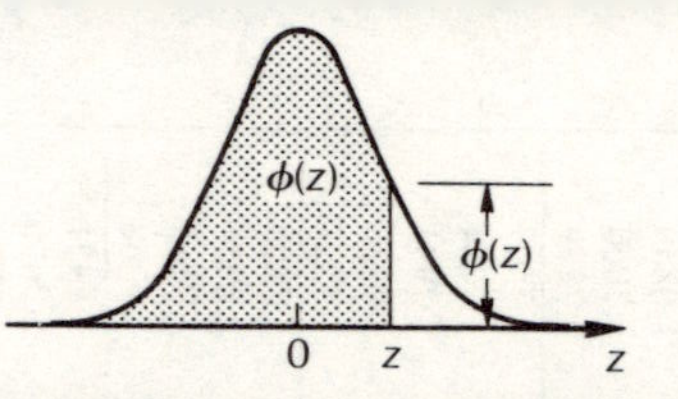

$\phi(z)$	z	0	1	2	3	4	5	6	7	8	9
0.3989	**0.0**	0.5000	5040	5080	5120	5160	5199	5239	5279	5319	5359
0.3970	**0.1**	0.5398	5438	5478	5517	5557	5596	5636	5675	5714	5753
0.3910	**0.2**	0.5793	5832	5871	5910	5948	5987	6026	6064	6103	6141
0.3814	**0.3**	0.6179	6217	6255	6293	6331	6368	6406	6443	6480	6517
0.3683	**0.4**	0.6554	6591	6628	6664	6700	6736	6772	6808	6844	6879
0.3521	**0.5**	0.6915	6950	6985	7019	7054	7088	7123	7157	7190	7224
0.3332	**0.6**	0.7257	7291	7324	7357	7389	7422	7454	7486	7517	7549
0.3123	**0.7**	0.7580	7611	7642	7673	7704	7734	7764	7794	7823	7852
0.2897	**0.8**	0.7881	7910	7939	7967	7995	8023	8051	8078	8106	8133
0.2661	**0.9**	0.8159	8186	8212	8238	8264	8289	8315	8340	8365	8389
0.2420	**1.0**	0.8413	8438	8461	8485	8508	8531	8554	8577	8599	8621
0.2179	**1.1**	0.8643	8665	8686	8708	8729	8749	8770	8790	8810	8830
0.1942	**1.2**	0.8849	8869	8888	8907	8925	8944	8962	8980	8997	9015
0.1714	**1.3**	0.9032	9049	9066	9082	9099	9115	9131	9147	9162	9177
0.1497	**1.4**	0.9192	9207	9222	9236	9251	9265	9279	9292	9306	9319
0.1295	**1.5**	0.9332	9345	9357	9370	9382	9394	9406	9418	9429	9441
0.1109	**1.6**	0.9452	9463	9474	9484	9495	9505	9515	9525	9535	9545
0.0940	**1.7**	0.9554	9564	9573	9582	9591	9599	9608	9616	9625	9633
0.0790	**1.8**	0.9641	9649	9656	9664	9671	9678	9686	9693	9699	9706
0.0656	**1.9**	0.9713	9719	9726	9732	9738	9744	9750	9756	9761	9767
0.0540	**2.0**	0.9772	9778	9783	9788	9793	9798	9803	9808	9812	9817
0.0440	**2.1**	0.9821	9826	9830	9834	9838	9842	9846	9850	9854	9857
0.0355	**2.2**	0.9861	9864	9868	9871	9875	9878	9881	9884	9887	9890
0.0283	**2.3**	0.9893	9896	9898	9901	9904	9906	9909	9911	9913	9916
0.0224	**2.4**	0.99180	99202	99224	99245	99266	99286	99305	99324	99343	99361
0.0175	**2.5**	0.99379	99396	99413	99430	99446	99461	99477	99492	99506	99520
0.0136	**2.6**	0.99534	99547	99560	99573	99585	99598	99609	99621	99632	99643
0.0104	**2.7**	0.99653	99664	99674	99683	99693	99702	99711	99720	99728	99736
0.00792	**2.8**	0.99744	99752	99760	99767	99774	99781	99788	99795	99801	99807
0.00595	**2.9**	0.99813	99819	99825	99831	99836	99841	99846	99851	99856	99861
0.00443	**3.0**	0.99865	99869	99874	99878	99882	99886	99889	99893	99896	99900
$\phi(z)$	z	0	1	2	3	4	5	6	7	8	9

EXAMPLES: $\Phi(1.2) = \text{prob}(Z \leqslant 1.2) = 0.8849$; $\Phi(1.23) = 0.8907$; $\Phi(1.234) = 0.8914$; $\text{prob}(Z \geqslant 2.3) = \text{prob}(Z \leqslant -2.3) = \Phi(-2.3) = 0.0107$ (making use of the symmetry of the normal distribution); $\text{prob}(0.32 \leqslant Z \leqslant 1.43) = \Phi(1.43) - \Phi(0.32) = 0.9236 - 0.6255 = 0.2981$.

Other normal distributions may be dealt with by standardization, i.e. by subtracting the mean and dividing by the standard deviation. For example if X has the normal distribution with mean 10.0 and standard deviation 2.0, $\text{prob}(X \leqslant 12.5) = \text{prob}(Z \leqslant (12.5 - 10)/2) = \text{prob}(Z \leqslant 1.25) = \Phi(1.25) = 0.8944$.

Table 2 Percentage points of the normal distribution

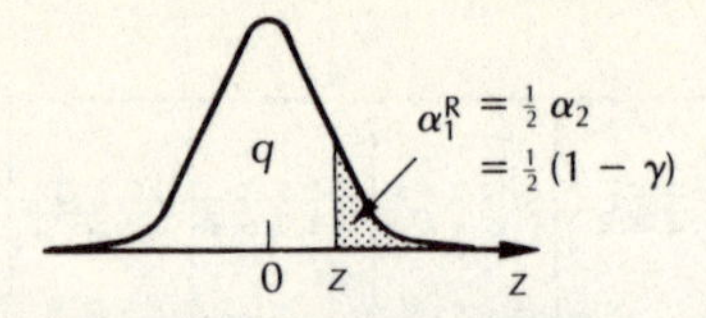

$q = \Phi(z)$	α_1^R	α_2	γ	z
0.50				0.0000
0.60	40%			0.2533
0.70	30%			0.5244
0.80	20%	40%	60%	0.8416
0.85	15%	30%	70%	1.0364
0.90	10%	20%	80%	1.2816
0.91	9%	18%	82%	1.3408
0.92	8%	16%	84%	1.4051
0.93	7%	14%	86%	1.4758
0.94	6%	12%	88%	1.5548
0.950	**5.0%**	10.0%	90.0%	1.6449
0.952	4.8%	9.6%	90.4%	1.6646
0.954	4.6%	9.2%	90.8%	1.6849
0.956	4.4%	8.8%	91.2%	1.7060
0.958	4.2%	8.4%	91.6%	1.7279
0.960	4.0%	8.0%	92.0%	1.7507
0.962	3.8%	7.6%	92.4%	1.7744
0.964	3.6%	7.2%	92.8%	1.7991
0.966	3.4%	6.8%	93.2%	1.8250
0.968	3.2%	6.4%	93.6%	1.8522
0.970	3.0%	6.0%	94.0%	1.8808
0.971	2.9%	5.8%	94.2%	1.8957
0.972	2.8%	5.6%	94.4%	1.9110
0.973	2.7%	5.4%	94.6%	1.9268
0.974	2.6%	5.2%	94.8%	1.9431
0.975	2.5%	**5.0%**	**95.0%**	1.9600
0.976	2.4%	4.8%	95.2%	1.9774
0.977	2.3%	4.6%	95.4%	1.9954
0.978	2.2%	4.4%	95.6%	2.0141
0.979	2.1%	4.2%	95.8%	2.0335
0.980	2.0%	4.0%	96.0%	2.0537
0.981	1.9%	3.8%	96.2%	2.0749
0.982	1.8%	3.6%	96.4%	2.0969
0.983	1.7%	3.4%	96.6%	2.1201
0.984	1.6%	3.2%	96.8%	2.1444
0.985	1.5%	3.0%	97.0%	2.1701
0.986	1.4%	2.8%	97.2%	2.1973
0.987	1.3%	2.6%	97.4%	2.2262
0.988	1.2%	2.4%	97.6%	2.2571
0.989	1.1%	2.2%	97.8%	2.2904
0.990	**1.0%**	2.0%	98.0%	2.3263
0.991	0.9%	1.8%	98.2%	2.3656
0.992	0.8%	1.6%	98.4%	2.4089
0.993	0.7%	1.4%	98.6%	2,4573
0.994	0.6%	1.2%	98.8%	2.5121
0.995	0.5%	**1.0%**	**99.0%**	2.5758
0.996	0.4%	0.8%	99.2%	2.6521
0.997	0.3%	0.6%	99.4%	2.7478
0.998	0.2%	0.4%	99.6%	2.8782
0.999	0.1%	0.2%	99.8%	3.0902

Notes to Table 2

The following notation is used in this and subsequent tables. q represents a quantile, i.e. q and the tabulated value z are related here by prob $(Z \leqslant z) = q = \Phi(z)$; e.g. $\Phi(1.9600) = q = 0.975$, where $z = 1.9600$. α_1, α_1^L and α_1^R denote significance levels for one-tailed or one-sided critical regions. Sometimes α_1^L and α_1^R values, corresponding to critical regions in the left-hand and right-hand tails, need to be tabulated separately; in other cases one may easily be obtained from the other. Here we have included only α_1^R, since α_1^L values are obtained using the symmetry of the normal distribution. Thus if a 5% critical region in the right-hand tail is required, we find the entry corresponding to $\alpha_1^R = 5\%$ and obtain $Z \geqslant 1.6449$. Had we required a 5% critical region in the left-hand tail it would have been $Z \leqslant -1.6449$. α_2 gives critical regions for two-sided tests; here $|Z| \geqslant 1.9600$ is the critical region for the two-sided test at the $\alpha_2 = 5\%$ significance level. Finally, γ indicates confidence levels for confidence intervals – so a 95% confidence interval here is derived from $|Z| \leqslant 1.9600$. For example with a large sample $X_1, X_2, \ldots, X_n$ we know that $(\bar{X} - \mu)(/s/\sqrt{n})$ has approximately a standard normal distribution, where $X = \Sigma X_i, n$ and the adjusted sample standard deviation s is given by $s = [\Sigma(X_i - \bar{X})^2/(n-1)^{1/2}$. So a 95% confidence interval for μ is derived from $|(\bar{X} - \mu)/(s/\sqrt{n})| \leqslant 1.9600$, which is equivalent to $\bar{X} - 1.96s/\sqrt{n} \leqslant \mu \leqslant \bar{X} + 1.96s/\sqrt{n}$.

Table 3 Percentage points of the Student t distribution

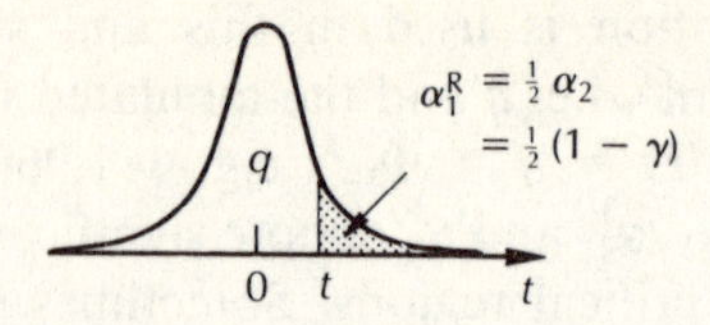

q	0.95	0.975	0.99	0.995
α^R_1	**5%**	$2\frac{1}{2}$%	**1%**	$\frac{1}{2}$%
α_2	10%	**5%**	2%	**1%**
γ	90%	**95%**	98%	**99%**
ν				
1	6.3138	12.7062	31.8205	63.6567
2	2.9200	4.3027	6.9646	9.9248
3	2.3534	3.1824	4.5407	5.8409
4	2.1318	2.7764	3.7469	4.6041
5	2.0150	2.5706	3.3649	4.0321
6	1.9432	2.4469	3.1427	3.7074
7	1.8946	2.3646	2.9980	3.4995
8	1.8595	2.3060	2.8965	3.3554
9	1.8331	2.2622	2.8214	3.2498
10	1.8125	2.2281	2.7638	3.1693
11	1.7959	2.2010	2.7181	3.1058
12	1.7823	2.1788	2.6810	3.0545
13	1.7709	2.1604	2.6503	3.0123
14	1.7613	2.1448	2.6245	2.9768
15	1.7531	2.1314	2.6025	2.9467
16	1.7459	2.1199	2.5835	2.9208
17	1.7396	2.1098	2.5669	2.8982
18	1.7341	2.1009	2.5524	2.8784
19	1.7291	2.0930	2.5395	2.8609
20	1.7247	2.0860	2.5280	2.8453
21	1.7207	2.0796	2.5176	2.8314
22	1.7171	2.0739	2.5083	2.8188
23	1.7139	2.0687	2.4999	2.8073
24	1.7109	2.0639	2.4922	2.7969
25	1.7081	2.0595	2.4851	2.7874
26	1.7056	2.0555	2.4786	2.7787
27	1.7033	2.0518	2.4727	2.7707
28	1.7011	2.0484	2.4671	2.7633
29	1.6991	2.0452	2.4620	2.7564
30	1.6973	2.0423	2.4573	2.7500
31	1.6955	2.0395	2.4528	2.7440
32	1.6939	2.0369	2.4487	2.7385
33	1.6924	2.0345	2.4448	2.7333
34	1.6909	2.0322	2.4411	2.7284
35	1.6896	2.0301	2.4377	2.7238
36	1.6883	2.0281	2.4345	2.7195
37	1.6871	2.0262	2.4314	2.7154
38	1.6860	2.0244	2.4286	2.7116
39	1.6849	2.0227	2.4258	2.7079
40	1.6839	2.0211	2.4233	2.7045

Notes to Table 3

The t distribution is mainly used for testing hypotheses and finding confidence intervals for means, given small samples from normal distributions. For a single sample, $(\bar{X} - \mu)/s/\sqrt{n})$ has the t distribution with $v = n - 1$ degrees of freedom (see notation above). So, e.g. if $n = 10$, giving $v = 9$, the $\gamma = 95\%$ confidence interval for μ is $\bar{X} - 2.2622s/\sqrt{10} \quad \mu \leqslant \bar{X} + 2.2622s/\sqrt{10}$. Given two samples of sizes n_1 and n_2, sample means $\bar{X}_1$ and $\bar{X}_2$, and adjusted sample standard deviations s_1 and s_2, $(\bar{X}_1 - \bar{X}_2)/\{s\sqrt{[(1/n_1) + (1/n_2)]}\}$ has the t distribution with $v = n_1 + n_2 - 2$ degrees of freedom, where $s = \{[(n_1 - 1)_1^2 + (n_2 - 1)s_2^2]/(n_1 + n_2 - 2)\}^{1/2}$. So if the population means are denoted μ_1 and μ_2, then to test $H_0{:}M_1 = M_2$ against $H_0{:}M_1 > M_2$ at the 5% level, given samples of sizes 6 and 10, the critical region is $(\bar{X}_1 - \bar{X}_1 - X_2)/(s\sqrt{1/6+1/10}) \geqslant 1.7613$, using $v = 6 + 10 - 2 = 14$ and α^{R}SB62;1 $= 5\%$. As with the normal distribution, symmetry shows that α_1^L values are just the α_1^R values prefixed with a minus sign.

Table 4 Percentage points of the chi-squared (χ^2) **distribution**

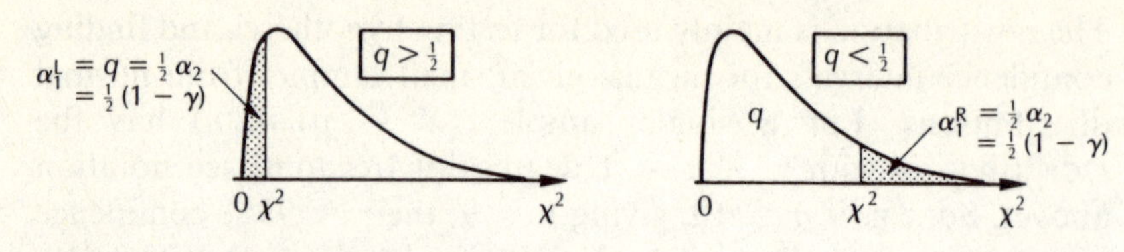

q	0.005	0.01	0.025	0.05	0.10	0.50	0.90	0.95	0.975	0.99	0.995
α_1^L	½%	1%	2½%	5%	10%						
α_1^R							10%	5%	2½%	1%	½%
α_2	1%	2%	5%	10%	20%		20%	10%	5%	2%	1%
γ	99%	98%	95%	90%	80%		80%	90%	95%	98%	99%
ν											
1	.00004	.00016	.00098	.00393	.0158	0.455	2.706	3.841	5.024	6.635	7.879
2	.0100	.0201	.0506	0.103	0.211	1.386	4.605	5.991	7.378	9.210	10.597
3	.0717	0.115	0.216	0.352	0.584	2.366	6.251	7.815	9.348	11.345	12.838
4	0.207	0.297	0.484	0.711	1.064	3.357	7.779	9.488	11.143	13.277	14.860
5	0.412	0.554	0.831	1.145	1.610	4.351	9.236	11.070	12.833	15.086	16.750
6	0.676	0.872	1.237	1.635	2.204	6.348	10.645	12.592	14.449	16.812	18.548
7	0.989	1.239	1.690	2.167	2.833	6.346	12.017	14.067	16.013	18.475	20.278
8	1.344	1.646	2.180	2.733	3.490	7.344	13.362	15.507	17.535	20.090	21.955
9	1.735	2.088	2.700	3.325	4.168	8.343	14.684	16.919	19.023	21.666	23.589
10	2.156	2.558	3.247	3.940	4.865	9.342	15.987	18.307	20.483	23.209	25.188
11	2.603	3.053	3.816	4.575	5.578	10.341	17.275	19.675	21.920	24.725	26.757
12	3.074	3.571	4.404	5.226	6.304	11.340	18.549	21.026	23.337	26.217	28.300
13	3.565	4.107	5.009	5.892	7.042	12.340	19.812	22.362	24.736	27.688	29.819
14	4.075	4.660	5.629	6.571	7.790	13.339	21.064	23.685	26.119	29.141	31.319
15	4.601	5.229	6.262	7.261	8.547	14.339	22.307	24.996	27.488	30.578	32.801
16	5.142	5.812	6.908	7.962	9.312	15.338	23.542	26.296	28.845	32.000	34.267
17	5.697	6.408	7.564	8.672	10.085	16.338	24.769	27.587	30.191	33.409	35.718
18	6.265	7.015	8.231	9.390	10.865	17.338	25.989	28.869	31.526	34.805	37.156
19	6.844	7.633	8.907	10.117	11.651	18.338	27.204	30.144	32.852	36.191	38.582
20	7.434	8.260	9.561	10.851	12.443	19.337	28.412	31.410	34.170	37.566	39.997
21	8.034	8.897	10.283	11.591	13.240	20.337	29.615	32.671	35.479	38.932	41.401
22	8.643	9.542	10.982	12.338	14.041	21.337	30.813	33.924	36.781	40.289	42.796
23	9.260	10.196	11.689	13.091	14.848	22.337	32.007	35.172	38.076	41.638	44.181
24	9.886	10.856	12.401	13.848	15.659	23.337	33.196	36.415	39.364	42.980	45.559
25	10.520	11.524	13.120	14.611	16.473	24.337	34.382	37.652	40.646	44.314	46.928
26	11.160	12.198	13.844	15.379	17.292	25.336	35.563	38.885	41.923	46.642	48.290
27	11.808	12.879	14.573	16.151	18.114	26.336	36.741	40.113	43.195	46.963	49.645
28	12.461	13.565	15.308	16.928	18.939	27.336	37.916	41.337	44.461	48.278	50.993
29	13.121	14.256	16.047	17.708	19.768	28.336	39.087	42.557	45.722	49.588	52.336
30	13.787	14.953	16.791	18.493	20.599	29.336	40.256	43.773	46.979	50.892	53.672
31	14.458	15.655	17.539	19.281	21.434	30.336	41.422	44.985	48.232	52.191	55.003
32	15.134	16.362	18.291	20.072	22.271	31.336	42.585	46.194	49.480	53.486	56.328
33	15.815	17.074	19.047	20.867	23.110	32.336	43.745	47.400	50.725	54.776	57.648
34	16.501	17.789	19.806	21.664	23.952	33.336	44.903	48.602	51.966	56.061	58.964
35	17.192	18.509	20.569	22.465	24.797	34.336	46.059	49.802	53.203	57.342	60.275

Notes to Table 4

The χ^2 (chi-squared) distribution is used in testing hypotheses and forming confidence intervals for the standard deviation σ and the variance σ^2 of a normal population. Given a random sample of size n, $\chi^2 = (n - 1)s/\sigma^2$ has the chi-squared distribution with $v = n - 1$ degrees of freedom (s is defined in the notes for Table 3). So if $n = 10$, giving $v = 9$, and the null hypothesis H_0 is $\sigma = 5$, 5% critical regions for testing against (a) H_1: $\sigma < 5$, (b) H_1: $\sigma > 5$ and (c) H_1: $\sigma \neq 5$ are (a) $9s^2/25 \leqslant 3.325$, (b) $9s^2/25 \geqslant 16.919$ and (c) $9s^2/25 \leqslant 2.700$ or $9s^2/25 \geqslant 19.023$, using significance levels (a) α_1^L, (b) α_1^R and (c) α_2 as appropriate. For example if $s^2 = 50.0$, this would result in rejection of H_0 in favour of H_1 at the 5% significance level in case (b) only. A $\gamma = 95\%$ confidence intervals for σ with these data is derived from $2.700 \leqslant (n - 1)s^2/\sigma^2 \leqslant 19.023$, i.e. $2.700 \leqslant 450.0^2 \leqslant 19.023$, which gives $450/19.023 \leqslant \sigma^2 \leqslant 450/2.700$ or, taking square roots, $4.864 \leqslant \sigma \leqslant 12.910$.

The χ^2 distribution also gives critical values for the familiar χ^2 goodness-of-fit tests and tests for association in contingency tables (cross-tabulations). A classification scheme is given such that any observation must fall into precisely one class. The data then consist of frequency counts and the statistic used is $\chi^2 = \Sigma\,(\text{Ob.} - \text{Ex.})^2/\text{Ex.}$, where the sum is over all the classes, *Ob.* denoting *ob*served frequencies and *Ex. ex*pected frequencies. These being calculated from the appropriate null hypothesis H_0. It is common to require that no expected frequencies be less than 5, and to regroup if necessary to achieve this. In goodness-of-fit tests, H_0 directly or indirectly specifies the probabilities of a random observation falling in each class. It is sometimes necessary to estimate population parameters (e.g. the mean and/or the standard deviation) to do this. The expected frequencies are these probabilities multiplied by the sample size. The number of degrees of freedom v = (the number of classes − 1 − the number of population parameters which have to be estimated). With contingency tables, H_0 is the hypothesis of no association between the classification schemes by rows and by columns, the expected frequency in any cell is (its row's subtotal) × (its column's subtotal) ÷ (total number of observations), and the number of degrees of freedom v is (number of rows − 1) × (number of columns − 1).

In all these cases, it is *large* values of χ^2 which are significant, so critical regions are of the form $\chi^2 \geqslant$ *tabulated* value, using α_1^R significance levels.

Table 5 Critical values for Spearman's rank correlation coefficient

$$r_s = 1 - \frac{6D^2}{n^3 - n}$$

α_1^R	5%	$2\frac{1}{2}$%	1%	$\frac{1}{2}$%
α_2	10%	5%	2%	1%
n				
1	–	–	–	–
2	–	–	–	–
3	–	–	–	–
4	1.0000	–	–	–
5	0.9000	1.0000	1.0000	–
6	0.8286	0.8857	0.9429	1.0000
7	0.7143	0.7857	0.8929	0.9286
8	0.6429	0.7381	0.8333	0.8810
9	0.6000	0.7000	0.7833	0.8333
10	0.5636	0.6485	0.7455	0.7939
11	0.5364	0.6182	0.7091	0.7545
12	0.5035	0.5874	0.6783	0.7273
13	0.4835	0.5604	0.6484	0.7033
14	0.4637	0.5385	0.6264	0.6791
15	0.4464	0.5214	0.6036	0.6536
16	0.4294	0.5029	0.5824	0.6353
17	0.4142	0.4877	0.5662	0.6176
18	0.4014	0.4716	0.5501	0.5996
19	0.3912	0.4596	0.5351	0.5842
20	0.3805	0.4466	0.5218	0.5699
21	0.3701	0.4364	0.5091	0.5558
22	0.3608	0.4252	0.4975	0.5438
23	0.3528	0.4160	0.4862	0.5316
24	0.3443	0.4070	0.4757	0.5209
25	0.3369	0.3977	0.4662	0.5108
26	0.3306	0.3901	0.4571	0.5009
27	0.3242	0.3828	0.4487	0.4915
28	0.3180	0.3755	0.4401	0.4828
29	0.3118	0.3685	0.4325	0.4749
30	0.3063	0.3624	0.4251	0.4670
31	0.3012	0.3560	0.4185	0.4593
32	0.2962	0.3504	0.4117	0.4523
33	0.2914	0.3449	0.4054	0.4455
34	0.2871	0.3396	0.3995	0.4390
35	0.2829	0.3347	0.3936	0.4328
36	0.2788	0.3300	0.3882	0.4268
37	0.2748	0.3253	0.3829	0.4211
38	0.2710	0.3209	0.3778	0.4155
39	0.2674	0.3168	0.3729	0.4103
40	0.2640	0.3128	0.3681	0.4051
41	0.2606	0.3087	0.3636	0.4002
42	0.2574	0.3051	0.3594	0.3955
43	0.2543	0.3014	0.3550	0.3908
44	0.2513	0.2978	0.3511	0.3865
45	0.2484	0.2945	0.3470	0.3822
46	0.2456	0.2913	0.3433	0.3781
47	0.2429	0.2880	0.3396	0.3741
48	0.2403	0.2850	0.3361	0.3702
49	0.2378	0.2820	0.3326	0.3664
50	0.2353	0.2791	0.3293	0.3628

References

Adamson, Peter 1991: Why reading keeps children alive. In N. Sherratt and I. Crosher (eds) Open University D103 Preparatory Pack. Milton Keynes: Open University Press.

DHSS 1980: *Inequalities in Health* (chairman Sir Douglas Black). London: DHSS.

Durkheim, Emile 1952: *Suicide, a Study in Sociology*. London: Routledge and Kegan Paul.

Caldwell, J. C. 1981: Maternal education as a factor in child mortality. *World Health Forum*, 2 (1), 76.

CSO 1980: *Annual Abstract of Statistics*. London: HMSO.

Erikson, B. and Nosanchuk, T. A. 1992: *Understanding Data*. Milton Keynes: Open University Press.

Foster, Kate 1989: *A Survey of Funeral Arrangements in 1987*. London: OPCS, HMSO.

Grant, J. 1992: The State of the World's Children. Oxford UNICEF, Oxford University Press.

Howell, David 1985: *Fundamental Statistics for the Behavioural Sciences*. Boston, MA: Duxbury.

Kelly, Gabrielle 1992: Robust regression estimators. *The Statistician*, 41 (3), 303–14.

Knight, Ian 1984: *The Height and Weight of Adults in Great Britain*. London: OPCS, HMSO.

Macfarlane, Alison and Mugford, Miranda 1984: *Birth Counts: Statistics of Pregnancy and Childbirth*. London: OPCS, HMSO.

Marsh, Catherine 1988: *Exploring Data*. Cambridge: Polity.

Maxwell, James 1953: *The Social Implications of the 1947 Scottish Mental Survey*. London: University of London Press.

Minitab Quick Reference 1991: Release 8, PC Version, Pennsylvania State College, PA: Minitab Inc.

Minitab Reference Manual 1991: Release 8, PC Version. Pennsylvania State College, PA: Minitab Inc.

Neave, Henry R. 1992: *Elementary Statistical Tables*. London: Routledge.

Open University 1987, Statistical methods, M345, Unit 9, Milton Keynes.

Population Trends, 69 (Autumn 1992). London: HMSO.

The Registrar Generals Decennial Supplement, England and Wales 1978. London: HMSO.

Rowntree, Derek 1991: *Statistics without Tears*. Harmondsworth: Penguin.

Ryan, B. F., Joiner, B. L. and Ryan, T. A. 1985: *Minitab Handbook*. Boston, MA: PWS-Kent.

Social Trends, 11 (1981). London: HMSO.

Tukey, J. W. 1977: *Exploratory Data Analysis*. Reading, MA: Addison-Wesley.

World Health Organization 1988: *World Health Statistics Annual 1988*. Geneva: World Health Organization.

Young, Paul T. 1966: *Motivation and Emotion*. New York: Wiley.

Index